AN ARCHITECT'S GUIDE TO CONSTRUCTION

TALES FROM THE TRENCHES

BOOK 1

AN ARCHITECT'S GUIDE TO CONSTRUCTION TALES FROM THE TRENCHES
BOOK 1

BRIAN PALMQUIST, ARCHITECT
AIBC, MRAIC, INTL. ASSOC. AIA, BEP, CP, LEED AP

First published 2015
by Quality-by-Design Software Ltd.
www.quality-works.net

3814 West 19th Ave., Vancouver, BC Canada V6S 1C8

Trade paperback ISBN 978-0-9939876-0-1
Kindle E-book ISBN 978-0-9939876-1-8
Epub E-book ISBN 978-0-9939876-2-5

Cover and interior design by Andrew Palmquist

Library and Archives Canada Cataloguing in Publication

Palmquist, Brian, 1951- author
An architect's guide to construction / by Brian Palmquist, Architect AIBC,
MRAIC, Intl. Assoc. AIA , BEP, CP, LEED AP.

Includes bibliographical references. Contents: Book 1. Tales from the
trenches.
Issued in print and electronic formats.
ISBN 978-0-9939876-0-1 (book 1 : pbk.).
ISBN 978-0-9939876-1-8 (book 1 : kindle)
ISBN 978-0-9939876-2-5 (book 1 : epub)

1. Building. I. Title.

TH145.P34 2015   690   C2015-900744-5
                   C2015-900745-3

Printed by IngramSpark
www.ingramspark.com

This book is for my children Andrew & Sarah
whose generation needs it.

Dedicated to my wife Linda
who encouraged me to write it all down.

It's sometimes hard to imagine that designers and builders can have significant impact in to-day's world. I believe we can, in three ways:

Firstly, by synthesizing materials and systems into efficient, functional and up-lifting structures for habitation, commerce and congress. Surely that is the fundamental purpose of design and construction?

Secondly, by using the most efficient and sustainable methods and systems to support the first goal, including the systems and approaches used to construct our designs.

Thirdly, by teaching our successors about the first two.

An Architect's Guide to Construction (AAGC) is designed to pass on to our successors the aspects of construction I have found to be both timeless and current. AAGC blends traditional words (albeit in an e-book format) with the most intellectually efficient environment ever known, the worldwide web. I urge you to explore the web environment created as a complement to these words, and to add your own experience either in the form of contributions to the sample architectural project available to readers of this e-book, or more conventionally as emailed comments or posts either to the LinkedIn group "An Architect's Guide", or to my publishers, Amazon.com and IngramSpark.com.

The success of any book rests with its readers. If you like AAGC, please let other design and construction colleagues know. Review and (hopefully) recommend it wherever you go for your books and e-books, including libraries and bookstores. The power of recommendation remains huge in the book and e-book world. I have begun to collect tales about business and quality that I believe will assist the next generation of designers and builders. I will he happy to share them if there appears to be an interested audience.

Writing An Architect's Guide to Construction has been an incredibly satisfying way to "give back". The financial rewards from publishing may be paltry, but if even one reader's design and construction life is improved by reading An Architect's Guide to Construction, then my reward will be huge.

As with most efforts, many individuals contributed directly or indirectly to this book. I have to start with some of the architects who mentored me – I consider I was lucky in that regard. The firm of Bain Burroughs Hanson Raimet was my first west coast employer. Ron Bain remains a particularly thoughtful architect who introduced me to a wide range of tasks and approaches. Howard|Yano Architects worked me hard but paid me well and fairly and taught me much; they also gave me my first taste of managing projects, which has led me to where I am to-day. Rick Hulbert taught me about concept planning and urban design as I worked for him as managing architect on two of Vancouver's seminal projects, Concord Pacific Place and Coal Harbour. He also taught me more about marketing professional services than anyone before or since.

After I opened my own practice, I continued to benefit from architectural colleagues because I often worked with them as a specialist building code or building envelope consultant. They are too many to name, but I always appreciated them and have enjoyed many repeat architectural and client relationships, which is the basis of any successful business.

Arc Rajtar first worked with me more than 12 years ago when I was struggling to create an ISO-compliant quality management system in order to preserve my insurability. In the face of the "leaky condo crisis", around the year 2000, the insurers of professionals in my home province were threatening to uninsure every architect and engineer. My insurer tipped me off to this danger, and suggested ISO 9001 certification might preserve my insurability, hence my ability to practice. After agreeing with me that my small firm could not manage the ISO workload using conventional paper, Arc recommended Mark Silvester as an ISO auditor/ software developer to work with. I told Mark that I would describe how to design and build a building and he would help me do so in an ISO-compliant fashion, but without the jargon. After ten years, we continue to debate the jargon!

Robert Campbell worked with Mark to maintain the integrity and security of the emergent application, Quality-Works.net (QW), so successfully that we have never lost even one byte of data in 10 years. He and his partner Cailin Green are also the talented singers and songwriters Campbell + Green now living in Nova Scotia. The magic of the 24/7/365 Internet allows Robert to continue his technical support from the other side of Canada.

Any book, especially a nonfiction work, benefits from colleagues who volunteer to act as "beta readers" – individuals who offer to read the book in manuscript form and provide thoughtful comments and suggestions. I am fortunate in having a significant number of colleagues "volunteer", including (in first name alphabetical order the way contractors like it):

Anita McReynolds-Lidbury, Chair-Elect, American Society for Quality Design and Construction Division, Certified Quality Auditor; Arc Rajtar, Managing Partner at Quexx International Ltd. + Quality Assurance Specialist, Levelton Consultants Ltd.; Doran Sharman, Vice President, Risk Management, Ledcor Industries Inc.; Jeff McLellan, Professional Services Division, Architects & Engineers Practice Leader (Western Region), BFL Canada; John Gresko, AIA, Senior Architect, HDR, Chicago; John Hackett, OAA, FRAIC, Vice President, Practice Risk Management, Prodemnity Insurance Company; Lawrence Bicknell, Vice President Professional Services National Practice Leader at BFL CANADA; Maura Gatensby Architect AIBC, Director of Professional Services, Architectural Institute of BC; Michael Ernest Architect AIBC, Executive Director (Retired), Architectural Institute of BC; Murray Mackinnon Architect AIBC, Vice President Sustainability, Ledcor Industries Inc.; Pierre Gallant Architect AIBC, AAA, FRAIC, Vice President, Morrison Hershfield Ltd.; Sean Rodrigues Architect AIBC, MRAIC, Director, Real Estate Development, Gracorp Capital Advisors Ltd. and Board Member, Canadian Architectural Certification Board (CACB); Susanna Houwen Architect AIBC, LEED AP, Principal at Susanna Houwen Architect.

While I have appreciated the comments of all, I must particularly thank: Murray Mackinnon, who is also an architect turned builder and provided balance where I sometimes missed it in earlier manuscripts – and is a fantastic copy editor; Pierre Gallant, whose comprehensive knowledge of the Canadian construction contract landscape through his involvement at CCDC ensured that contract references, especially Tale #5, are current and accurate; and Susanna Houwen, who added the perspective of a younger architect I respect as she carefully read every tale and commented on most of them.

This book is built upon the "Construction Administration" course I have been teaching to Intern Architects in British Columbia since 2004. When I started teaching I shared the material and delivery with Michael Ernest. Michael always told tales to explain the content – a few are included in this book because they remain current. When Michael became Executive Director of the Architectural Institute of BC he was unable to continue teaching, so I carried on by myself. But I would have been unable to do so without the support of the Institute's staff, in particular the Professional Development Coordinator, Aleta Cho, who put up with my continuous course refinements and other shenanigans until she retired recently after 30 years service at the Institute. The dedication of Aleta and her colleagues has been essential to my modest efforts and the more meaningful contributions of the many architects who look after the public interest in the pursuit of architecture.

My tales from the trenches are all true stories that help underline challenging aspects of design and construction not generally taught in schools or offices. I have had to make them anonymous to protect the innocent (and guilty). A few readers may recognize a specific tale in which they played a part. Whether or not you recognize your part, I thank you for your teaching and your patience with my lifelong learning.

THANKS FOR READING

"You know about this stuff. If you're not happy, then you write it down!"

I had just returned from an architectural continuing education conference – you know, the kind you go to every year or so to get enough continuing education points to satisfy the profession that you are actually paying attention to what's going on in the world. I often present at such occasions, but this year my *topic de jour* was out of favor, so for the first time in some time I attended just to listen (and top up my own points).

Being an attendee rather than a presenter affords a very different view of the design and construction world. Here's what I noticed:

For the first time we western Canadians were co-sponsoring our conference with the American Institute of Architects (AIA) chapter just south of the border. It felt like there was a higher proportion of younger architects from the US present than from Canada.

Perhaps the numbers of young US architects seemed higher because they were, on average, a bit more strident, a bit more desperate than their Canadian counterparts.

As the conference progressed, I had the occasion to attend a handful of business-oriented presentations. At these events, questions from the younger audience became increasingly pointed and probing. It was clear that pretty pictures with cursory business facts were just not cutting it. The audience wanted clear direction on a variety of practical subjects that were not on the conference agenda.

Listening more carefully to presentation content and questions arising, it became clear to me that the younger cohort was not being well served. When I complained to my wife, she responded with this book's opening remark.

I decided I might be able to help in some modest way. I have been teaching construction contract administration on a periodic basis to student and intern architects for 18 of the last 24 years – some might say that makes me part of the problem rather than the solution! Typical sessions are a half or full day – that's what the professional associations allot, not nearly enough time in my opinion, and evident in the frustrations of younger conference attendees.

During the years I have been teaching, I have also developed a commercial cloud-based project and practice management application for use in design and construction practice, into which I have packed my accumulated knowledge and the occasional sprig of wisdom, as well as making project and practice management much more quickly and efficiently available.

1

This e-book is designed to compress what has taken me years to learn and capture to the cloud into an e-book format that delivers focused content by hyperlinking the table of contents to e-book and other content. The book started as a couple dozen chapters, but has been recast as 70 shorter stories, "tales" encompassing the most challenging aspects of the construction phase for designers and builders.

By delivering content in e-book format, and also hyperlinking that e-book content to instructive YouTube videos, external web sites and my web-based project management content, I am hoping to create a resource that readers can return to for answers to the questions that arise during the course of their design and construction work. The e-book format also allows me to deliver the product at a fraction of hard copy cost, an important consideration for anyone with a limited budget. (There is also a paperback copy option for those who prefer, but it does cost more).

The Purpose of this Book

This book will help free you and your project teams to design and build by focusing on two aspects of the building construction phase: the critical knowledge and experience gaps that are not generally taught in design schools or apprenticeship programs; and just as important, the most efficient means to apply that knowledge and experience to your projects. This book will help whether you are a sole practitioner working with "mom and pop" contractors, or a multinational, multi-branch organization specializing in design-build or P3. It will help younger architects and construction professionals build more successfully by managing construction more efficiently. It will help older professionals fill in their construction phase knowledge gaps and sharpen their skills. It will expose designers and builders both to the other's perspective.

Design and construction professionals of all ages will learn how to become more efficient in their work, simplifying their approach by considering four universal principles for design and construction project management, supported by just three integrated web-based "toolsets". If you choose not to use the recommended tools or principles, you will at least learn the means to evaluate the many offerings in the marketplace. You will become more profitable during the construction phase of a project. You will have more time to take on more work, spend more time with your family, or pursue your other passions.

The "take on more work" option will help you find and keep work in to-day's tough economy, while executing it more efficiently, hence profitably. This is increasingly important because the past five years have been very challenging for designers in many countries. In his 2010 book "Down Detour Road: An Architect in Search of Practice"[1] , American author and graduate architect Eric Cesal notes that in the United Kingdom from 2008-2009, while the general unemployment rate doubled, there was "…nearly an eightfold increase in the

1   Cesal, Eric J., Down Detour Road: An Architect in Search of Practice, MIT Press, Cambridge, MA, USA, 2010.

number of unemployed architects from just the prior year." The Economist magazine estimates that between 2008 and 2012 "…demand for [architect's] services shrank by as much as 40%, the RIBA reckons."[2] In his more recent book "Architecture 3.0 – The Disruptive Design Practice Handbook"[3] author Cliff Moser AIA notes that the American Institute of Architects has identified that the number of employees in US architecture offices has declined by 25 percent since 2008.[4] And in Canada, perhaps somewhat less affected by the global recession than the US or the UK, the 2013 survey on behalf of the Canadian Architectural Licensing Authorities (CALA), which represents architectural licensing authorities across Canada, has identified that intern architects and their employers are unhappy about preparation for practice, especially the construction phase.[5]

Regarding the general satisfaction of intern architects and their employers across all Canadian architectural internship programs, The CALA survey concluded this: "…having conducted several hundred other surveys frequently for membership-based organizations, this is a failing grade."

Figure 1 - The Czar's Architect on a Pedestal - Going, going…[6]

2 The Stirling prize goes to the Everyman theatre", The Economist magazine. October 16th 2014, http://www.economist.com/node/21625925/print
3 Moser, Cliff, Architecture 3.0 – The Disruptive Design Practice Handbook, Routledge, New York, 2014, page 11.
4 Moser, Page 12, Figure 1.1
5 Framework Partners Inc., "2013 National CALA (Canadian Architectural Licensing Authorities) Survey", December 2013
6 Photograph of the bust of the Architect of the Church of the Bleeding Heart in St. Petersburg, Russia, taken by the author 2014.

Meanwhile, Moser proposes a radical reconsideration of what architects do and how they practice, based upon discarding traditional practice and adopting a new *"...fundamental role of "design for solutions."* He identifies *"...design for buildings as a separate specialized activity (but not the core activity) that the architect may or may not choose to practice."*[1] Moser's opinion seems supported by Figure 3 below.

I am not entirely in agreement with Cliff. Ever the optimist, I notice that concurrent with our AIA, RIBA and CALA-measured angst, the emergence of tablet computers and powerful browser-based software has finally created a new, more efficient environment for professional involvement in construction.

I am of the belief that traditional architectural practice can not only be saved, it can be rejuvenated by the value of our collective experience empowered by emergent tablet and smartphone technology.

This book is about that experience and empowerment. My goal is to free the project team to design and build.

Who Should read this Book?

If you are studying design and construction, or relatively new to it, you should read this book from cover to cover, as it is written in approximate project flow order and the e-book Table of Contents links let you jump around quite effectively. The e-book's workflow order means that you can jump straight to a specific tale when you are looking for focused knowledge. If you are more seasoned, just read this Introduction, then skip straight to the tales that describe your biggest challenges.

Be patient if you read information that you believe to be self-evident. Teaching designers and builders both, I am reminded every day of the very different perspectives each brings to a common activity. Designers assume builders understand design intentions – "What I intended was..." or "The design concept is..." are swear words for builders, as they are frequently followed by information that is not at all obvious to anyone BUT the designer.

Conversely, builders assume designers know how to design buildings that are easy to construct, so why don't they? "Where can I possibly put the crane?...", "There's no room for delivery and laydown...", "That detail is unbuildable..." are three of the phrases that come to mind.

---

1   Moser, Page 4.

Figure 2 - Builder site plan different than Designer site plan

Also be patient, AAGC is written for an audience that includes students, younger designers and builders as well as some seasoned practitioners.

Over the past several years I have worked on almost every possible project type and scale. I have mined this experience for the 70 tales that follow. I play no favorites. Every reader will disagree with some of what AAGC says. I encourage you to voice your disagreements and other contributions either as notes in the sample project you will have free access to, comments on the LinkedIn group "An Architect's Guide", comments or preferably reviews with my e-book publisher, Amazon/Kindle. Or of course by direct email to me at bpalmquist@quality-works.net .

Three Times Four Makes Two

There are seventy tales in this e-book, leading me to ask, *"How did the simple master builder profession that dominated up until about the year 1800 become the complexity that we are so often faced with to-day? Can we simplify it?"* I say yes.

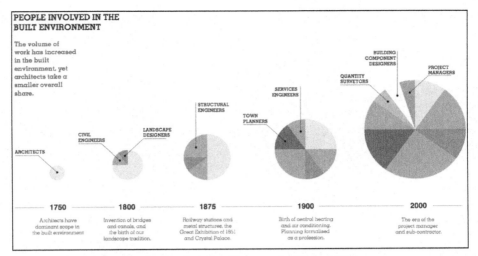

Figure 3 – Smaller Pieces of a Bigger Pie[1]

In the post-industrial, pre-digital era, managing just the construction phase of even a moderately complex building project involves at least 35 steps/functions requiring almost as many logs and at least 850 individually prepared forms, templates and reports. In the table next page, "Log" means a log is usually prepared; "D" involves the Designer; "B" involves the Builder; "C" involves the Client; "Qty" is a typical quantity based on a 12-month construction duration with excellent bid documents, 5 bidders, monthly site meetings and only monthly field review reports from key consultants; other issue quantities are optimistic based on my experience:

---

1   The Farrell Review Team, "Our Future in Place", The Farrell Review of Architecture + The Built Environment, Conclusions and Recommendations, Department of Culture, Media & Sport, UK, p. 63

	From	Log	D	B	C	Qty
01	**Bidder list**	■	■	□	□	1
02	**Project risks**	■	■	■	□	2
03	**Addenda**	■	■	□	□	4
04	**Bid form**	□	■	■	■	5
05	**Award letter**	□	□	□	■	1
06	**No-award letter**	□	□	□	■	1
07	**Contract breakdown**	□	□	■	□	1
08	**Value engineering**	■	■	■	□	5
09	**Monthly invoice**	■	□	■	■	12
10	**Payment certificate**	■	■	□	■	12
11	**Statutory Declaration**	□	□	■	□	11
12	**Project directory**	■	□	□	□	1

	From	Log	D	B	C	Qty
13	Schedule	☐	☐	■	☐	1
14	Work breakdown structure	☐	■	■	☐	2
15	Project quality plan	☐	■	■	☐	2
16	Site meeting minutes	■	☐	■	☐	12
17	Requests for Information	■	☐	■	☐	10
18	Requests for change/quotation	■	☐	■	☐	10
19	Site Instructions	■	■	☐	☐	10
20	Contemplated Change Orders	■	■	☐	■	10
21	Change Orders	■	■	☐	■	10
22	Change Directives	■	■	☐	■	4
23	Submittals	■	■	■	☐	20
24	Mockups	■	■	■	■	10

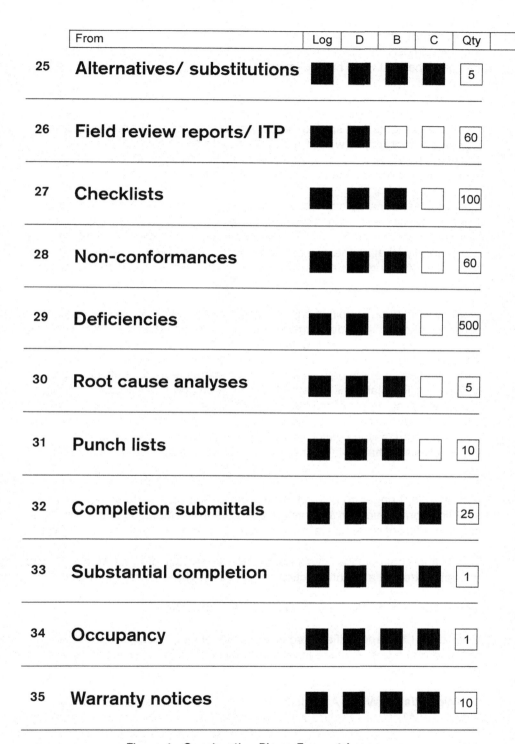

From	Log	D	B	C	Qty
25 Alternatives/ substitutions	■	■	■	■	5
26 Field review reports/ ITP	■	■	☐	☐	60
27 Checklists	■	■	■	☐	100
28 Non-conformances	■	■	■	☐	60
29 Deficiencies	■	■	■	☐	500
30 Root cause analyses	■	■	■	☐	5
31 Punch lists	■	■	■	☐	10
32 Completion submittals	■	■	■	■	25
33 Substantial completion	■	■	■	■	1
34 Occupancy	■	■	■	■	1
35 Warranty notices	■	■	■	■	10

Figure 4 - Construction Phase Forms & Logs

In wrestling with this alphabet soup of documentation, I struggled for many years to simplify the hand I had been dealt as a designer, later builder. The advent of the digital era helped immeasurably.

What started as tables, logs and forms typewritten or drafted onto small Mylar sheets morphed into the next generation of standard office tools, word-processed documents and spreadsheets. As computers proliferated, specialized schedules emerged, as well as the ability to attach data first to two-dimensional CADD, now three and four-dimensional BIM models. But most of the 850 documents I mentioned in the table above remained stuck in MSOffice.

As computers became more used and useful, the number of applications addressing design and construction functions grew rapidly, then exponentially with the introduction of tablets, smartphones and their "apps." And yet, we are still generally treating the communications content of the construction phase as 35 sets of documents each managed in its own fashion and communicated via generally uncoordinated email, occasionally via courier.

Working with "apps", applications and "the cloud", each individual designer and builder is becoming gradually more efficient in her/his personal experience of working in design and construction. But as the complexities of communicating between multiple platforms grow (what experts call interoperability) more and more time is spent developing and managing interfaces between different communication platforms, less and less time on creating and communicating the essential elements of design and construction. Think how much time you take each day to, first, "boot" your computer; then, each time you switch applications from, say, email to CADD to BIM to word processing, back to email, etc. Each such change in your operating environment is time consuming and tiring.

In response, I began to develop ever-simpler approaches to design and construction communication, characterized by flexible adaptation to the communication models of others by taking advantage of emergent web-based software.

Out of that continuing analysis have come three toolsets and four principles to capture design and construction – what I call "Three times four makes two."

Three Toolsets to Manage Design and Construction

Over time I have identified, organized and refined just three "toolsets" that seem to manage everything in design and construction with better speed and greater simplicity:

1. i-WorkPlan (i-WP) – A traditional work plan, also called a work breakdown structure (WBS), is simply a list of all the key processes and procedures involved in design and construction. References such as the Canadian Handbook of Practice for Architects (CHOP) contain static lists of steps, often without instructions for the novice:

II   Schematic Design Phase

Part A — *Tasks prior to starting the schematic design phase*

1.	**Obtain name of the client's authorized representative.**
2.	**Obtain the client's project brief.** Confirm the client's space needs and other program requirements.
3.	**Establish project filing system.**
4.	**Assign personnel to the project:** • project architect • designer • technical staff • other
5.	**Assemble and review all applicable requirements of Authorities Having Jurisdiction (e.g., site plan control, applicable zoning or land use, and code requirements).** Review with Authorities (refer to Chapter 1.2.4).
6.	**Establish project schedule, including completion dates for each phase of project.** Advise the client, staff, and all consultants.
7.	**Finalize Consultant Agreements** **Negotiate, prepare, and execute consultant agreements. When required, obtain the client's approval of consultants:** • structural • mechanical • electrical • other
	Request and receive from each consultant proof of professional liability insurance coverage: • structural

Figure 5 – Partial Work Breakdown Structure[1]

When I first began to work with quality management systems, I noticed three things about traditional work plans or WBS's: firstly, their steps seldom included instructions of value to less experienced practitioners; secondly, their adaptation to become project-specific quality manuals, work breakdown structures or project plans took way too long; finally, once developed, they were placed on shelves and seldom referred to on a day-to-day basis.

I realized that web-based technology could automate work plan preparation; also that including interactive work instructions, project diaries and supporting forms and templates into project work plans would eliminate shelf time because the plans would become helpful in day-to-day work.

The resulting i-WorkPlan describes an interactive web-based work plan that underlies each of my projects, comprising step by step procedures from design through construction that create the project roadmap. By mapping out the entire project at the beginning, I am much better equipped to deal with changes and the unexpected.

---

1   Hobbs, Jon, Editor-in-Chief, Canadian Handbook of Practice for Architects, 2nd edition, January 2009, page CH-33/5

Although familiar to builders and project managers as a work breakdown structure, the i-WP is unfamiliar to many designers and their biggest challenge in terms of new ways of thinking.

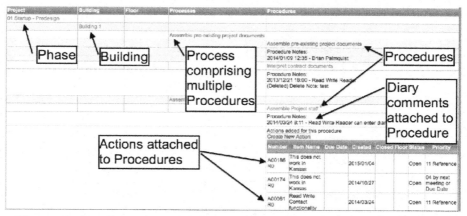

Figure 6 - Small portion of i-WP showing diary comments & attached i-Actions

For a project type that I have experience with, it takes me about ten minutes to create my i-WP, a few minutes more to customize it for a repeat client or a particular municipality.

Where my experience is more limited but I am confident in proceeding, I can quickly create a generic i-WorkPlan, then flesh it out as the project develops and I learn new things about its design and construction. For the airport project I have recently set up with an i-WP, it required about 2 hours (after preliminary meetings and discussions) to develop a functional i-WP encompassing 5 design and construction phases, 28 building zones and 129 floor segments, all within 50 overlapping evacuation zones. – more than 500 distinct tasks to consider, many with multiple actions.

As I use an i-WP, if new requirements emerge that are specific to, for example, the current project's client or community, I can add these to the i-WP while tagging them so that they appear automatically in future projects for that client or in that community. If I amend an existing procedure (for example when an existing client changes one of their procedures that I interact with), then that amendment is immediately deployed to every other current and future project. Thereafter, each time a new project is started with that client, its i-WP incorporates the amended Procedure. In this way, my clients and collaborators can see that my practice continuously learns from them. Not bad PR!

Over time, I have identified 10 categories of information that encompass most i-WP customization. I call these "the 10C's":

Name _____

01   **Clients** I work with
_____

02   **Contractors** I work with
_____

03   **Consultants** I work with
_____

04   **Contract Types** - lump sum, design build, P3, etc.
_____

05   **Cost range:** Small project vs. megaproject
_____

06   **Communities** I work in
_____

07   **Construction type:** Concrete, steel, wood, etc.
_____

08   **Calendar (schedule)** I work to: Fast track, etc.
_____

09   **Climate of the project:** Regional variations
_____

10   **Complexity of the project:** Warehouse versus airport
_____

Figure 7 - the 10C's

2. i-Action (i-A), the second toolset, describes the integrated set of transient communications that captures i-WP work product and deals with emergent issues – most of the 850+ specific forms, logs and templates noted in Figure 4 above. I-A's are attached to the i-WorkPlan while also informing practice knowledge.

The i-A tool set concept arose from my noticing that the wide variety of individual forms and templates used in design and construction is all "peas in a pod", sharing these seven characteristics:

1. Their **quantity** on a project is **unpredictable** in advance.

2. They usually **involve simple workflow** (e.g., architect to contractor back to architect, then to client). Historically, this has been captured as a "Traveler", a paper checklist/transmittal that accompanies a product or document as it moves through its workflow.

3. They are used by designers to **manage workflow** such as approval of changes to the work, also by builders to manage workflow such as product assembly:

Figure 8 - Traditional paper Traveller tucked into Curtain Wall

4. The workflow and nature of i-A's often **evolves** – an RFI becomes a Site Instruction, then perhaps a Contemplated Change and a Change Order.

5. They can originate from any of the members of a project team and **may be reassigned** "on the fly" as the i-A evolves.

6. Their default **communication vehicle** is via email, the antithesis of organized discourse.

7. Recipients may wish to view them on a **laptop, smartphone or tablet.**

The i-WP noted above will include in its body as attachments to specific tasks all of the i-A forms and templates needed to execute the project. As a project unfolds, some elements will scarcely be used, others such as submittal review may be used dozens of times, and will be assembled and managed through the logic of the i-WP.

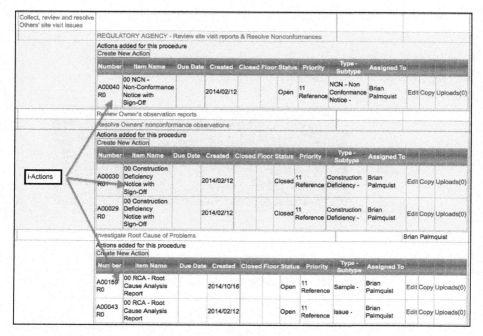

Figure 9 - i-A's added through i-WP

As with the i-WP, where an existing procedure form such as "Submittal" needs to change for Client A because of their internal requirements, the original "Submittal" i-A can be quickly amended to become "Client A Submittal" attached to the Procedure entitled "Client A Submittals." In addition to informing the balance of the current project, this amended procedure and i-A will be automatically included in other current and future projects for Client A.

3. i-KnowHow (i-KH), the third and final toolset, encompasses the experience, expertise and knowledge database that each designer and builder needs to assemble in order to evolve, become efficient and more competitive. It is sometimes called an operations manual, or "how we do things here." Each company's project i-WorkPlan evolution is constantly refined by the experiences of i-KnowHow.

In An Architect's Guide (AAG) i-KH approach, evaluation of a new product, for example, has already been captured in an i-A, including all the research, web links, back and forth, etc. That i-A can be placed in a knowledge "purgatory" that advises company Subject Matter Experts (SME's) that there is some possible new knowledge for their evaluation. Their evaluation may be straightforward or may involve additional communication, all captured in notes to the original submission.

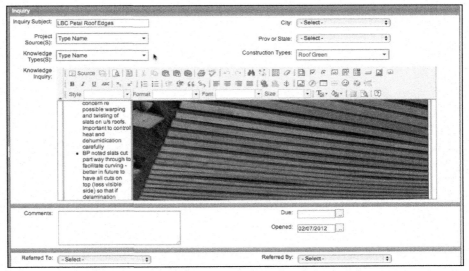

Figure 10 - i-KH proposed new knowledge

When the evaluation has been completed, the new knowledge can be suitably labeled (e.g., "Awesome new metal cladding" or "Metal Cladding to avoid") and moved from knowledge purgatory to the firm's proprietary knowledge database, accessible to its employees.

Now consider Project B for the same company, different project architect, and same material request. The new project architect, presumably unaware of previous research, can search the database and find if the firm has previously reviewed the proposed material and what the results were. The i-KH can be imported into Project B as an i-A and communicated to the new team, perhaps as *"…should work fine here…"* or *"….reviewed on another project, not approved by our firm for use on our projects for these reasons…."*

Figure 11 - i-KH before conversion to i-A

An i-KH captured as an i-Action in another project can be integrated into a current project with just a few clicks.

These tool sets support and clarify many of the most challenging aspects of construction. You may have better principles and tool sets, and if this book causes you to think about them, refine and clarify them, then you will add value to your own professional life. All good!

Refer to the LinkedIn Group "An Architect's Guide" for a more detailed discussion of these toolsets.

Four Principles to Manage Design & Construction

In my current role, to teach a broad audience of construction people about the practice of quality-managed construction, I have struggled with the complexities of quality management systems such as Kaizen, ISO 9001, Lean, Six Sigma and Total Quality Management. Out of these struggles have arisen four principles that demystify these complex management systems and are just as applicable to the practice of design as the practice of construction.

RECORD the journey could be paraphrased in traditional quality management words: "Say what you do, do it and record it." It identifies the need to list all of your design and construction procedures and provide instructions about how to perform them together with any forms or templates that may be needed. Contractors generally call this an operations manual. Designers generally call this "the way we've always done it." Where the list of procedures for a specific project is drawn from core design and construction knowledge, the result is a project-specific work plan – what AAGC earlier called an i-WorkPlan (i-WP).

RESOLVE the issues refers to the fact that every project has emergent challenges that need to be resolved systematically in order to keep the project moving and to build client, consultant and contractor partnerships. Many projects founder on a team's inability to resolve issues in a timely fashion. Construction lawyers all have their own tales of cases won or lost because an issue was or was not resolved and expeditiously closed.

REVIEW the results acknowledges the importance of continuous measurement so that project teams and management can identify how they are doing and have the ability to make improvements and refinements during the course of each project. Measurements that lead to improvements include: checklists that first guide design, then installation, and measure its successes and shortcomings; the incidence and resolution of non-conformances and

deficiencies during construction; and the day-to-day progression through the tasks included in a project's i-WP.

REMEMBER & learn to improve has been identified by design and construction researchers as the biggest process challenge in to-day's "just in time" design and construction environment.[1] It is all about identifying and capturing new or refined knowledge, then integrating it into the existing body of processes and procedures and deploying the new knowledge to all current and future projects.

Each of these four principles is challenging to learn and apply, especially in a pre-tablet/smartphone environment. Some of the tales in this book will show how technology makes the individual principles easier to work with.

Refer to the LinkedIn Group "An Architect's Guide" for a more detailed discussion of these four principles.

Why Bother Linking this e-book to the Internet?

Although e-books are relatively new, and the Internet relatively "old", the two still generally don't play well together. Amazon and others have done a remarkable job selling and disseminating e-books over the Internet, but once you "open" an e-book, it's pretty much disconnected from the Internet that sold it to you.

That disconnection may arise from the fact that early e-reader devices made by Kindle, Kobo and Barnes & Noble had only rudimentary connections to the Internet, useful only to download e-books. By comparison, since 2012, there are more tablet computers than e-readers. In response, most of the e-readers that are not tablet computers have now added web browsers[2]. Why not take advantage of that improvement to link e-book content to social media and other inhabitants of the web?

---

1   Tan, H.C. et al., Capture and Reuse of Project Knowledge in Construction, Wiley-Blackwell 2010, Sussex, UK, pp. 2-3.
2   Wikipedia, Comparison of e-book readers, en.wikipedia.org/wiki/Comparison_of_e-book_readers, indicates that of the major e-book readers, only Barnes & Noble's Nook does not include a web browser. By comparison, all noted Amazon and Kobo e-readers include a web browser. The Wikipedia chart indicates 25 of 41 noted e-readers include a web browser, 12 of 41 do not and 4 of 41 are "unknown."

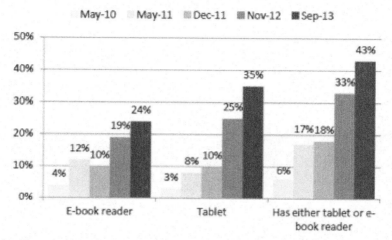

Tablet and e-reader ownership

% of Americans ages 16+ who own e-book readers, tablet computers, and at least one of those devices

Source: Most recent findings come from Pew Research Center Internet Project Library User survey. July 18-September 20, 2013. N= 6,224 Americans ages 16 and older. Interviews were conducted in English and Spanish and on landline and cell phones. Margin of error is +/- 1.4 percentage points for the total sample.

Note: The 2010 and 2011 surveys were conducted among those ages 18 and older.

Figure 12 - Tablets Overtaking E-book Readers[1]

Throughout this e-book I have inserted links to web resources. These are mostly links to external web sites, YouTube videos or to my own application. Where it is useful to demonstrate a procedure, I generally rely on simple YouTube videos made from my laptop computer. These will have URL's such as http://youtu.be/ttvRG6AHkIU. If the e-reader you are using to read this e-book supports connection to the Internet, you should be able to tap on the link and open the video. If you are reading a paper copy of this e-book, you will need to transpose the URL.[2]

Where I have found other web content that is helpful, I will include its URL, such as aibc.ca, riba.org, aia.org, etc. Suffice it to say I assume no liability for anything you read on external web sites.

Some other content may be best illustrated using my web-based project and quality management application, Quality-Works.net® (QW).

1   Pew Research Internet Project, "Tablet and E-reader Ownership Update", October 18, 2013, http://bit.ly/1DJeTtt

2   To assist with transcription of often long URL's for those reading a paper version of this e-book, I use a free application called Bitly.com to compress a hyperlink that starts as http://home.quality-works.net/Projects/TaskPlan.aspx?project=soy510CHcQNg=&page=Project%20Quality%20Plan#id3199 So that it ends up much shorter as http://bit.ly/1dc3aFa It requires an extra few moments for a Bitly URL to work, but far less than the time it takes to transcribe the long form versions.

Refer to Appendix #1 for further information regarding QW, including how to gain free access to the sample project that houses many of the illustrations in AAGC.

The Structure of this Book – Tales from the Trenches

In my experience, the most effective means to communicate the complexity of construction is by telling tales. In my 18 years of (part time) teaching and 35+ years of practice, I have come to understand that seemingly complex ideas are best taught by using brief stories to introduce construction challenges. This sets the stage for the introduction of what are often simple tools to get the job done.

Figure 13 - Telling Tales Works

I use tales almost every day whether I am talking with architects, engineers, owners or contractors. Students and audiences through the years have consistently rated my tales as most useful – not to say I am a great storyteller, just that this methodology works.

So we will introduce a series of construction phase tales in the most important areas where designers and builders may learn. Each tale is based upon real experiences. Names and places have been changed to protect the innocent and avoid me being sued!

Like the i-WorkPlans I use to describe each project, the tales are organized in approximate project workflow order and grouped by general subject. For example, tales about the importance and content of the first project site meeting are told well in advance; stories involving post occupancy warranty issues come last. And the tales relating to submittals are grouped together somewhere in the middle.

Tale #70 summarizes the content of the 69 preceding tales as ten rules that will guide you to a more pleasant and profitable work life.

The appendices to this book include a glossary of common terms as well as further information about collaborating and sharing our construction knowledge.

What this book is NOT

This is not a reference book encompassing all of the construction phase, which is why it in entitled "An Architect's Guide...", not "The Architect's Guide...". It focuses on those aspects of the construction phase of architectural practice that are most challenging, especially for younger practitioners. AAGC does not address business development, contracts or the design process itself – these may be the subjects of future "Tales from the Trenches" e-books.

Disclaimers

This book tries to help you make process, procedure, documentation and recording decisions that will allow you to practice design and construction better, especially the construction phase. But as the author, I am not responsible for your use or abuse of the contents of this book. My advice in this book should not be construed as professional or project-specific advice.

These tales are based on real experience, much of it mine but with a generous measure of others'. The contents do not represent the specific policies, procedures or professional practices of any design or construction professional body, my ten employers, five companies or two business partners since university. Opinions are mine. Any errors or omissions are mine.

There are a few locations in the book where I refer to specific products or services. With one exception, no one pays me directly or indirectly to mention or praise a product or service.

Starting in about 2000, I developed and own a commercially available web-based project and quality management application called Quality-Works.net (QW) so I do derive benefit from any subscriptions to it. I mention it periodically in this e-book because it uses the three tool sets in many ways, and provides useful illustrations.

Since Quality-Works includes blog content and readers' names may be visible, I reserve the right to edit or delete inappropriate commentary, restrict or terminate access to QW at my sole discretion.

70 Tales from the Trenches

Everyone likes a good story. Each of AAGC's "Tales from the Trenches" will generally have the following structure:

The title will try to introduce the concept being considered. It will include the point of the tale, not just the subject matter.

Where there is a quote that captures the starting point of the tale, it will be inserted next.

The balance of each tale generally has three parts:

What's the point? The quote will be followed by a summary of the story details and key issues raised. What is causing difficulty? What is wasting time and effort?

What are the principles & best practices? What are the key design and construction principles underlying the challenges of the tale? What is typical and what is best practice? Often this is not known to the designer/builder, or she/he is too timid to insist on it.

What is the simplest solution that works in the context of the entire project? AAGC will make recommendations on how to proceed based on experience.

Summary of Principles & Tools – To reinforce the relationship between each tale and the underlying four principles and three tool sets, each tale will end with a short summary of the referenced principles and tools. Not every principle and tool may be needed for each tale.

RECORD the journey	What records are needed?	**Summary Principles & Tools**	
RESOLVE the issues	Typical issues arising	**i-WorkPlan (i-WP)**	Specific procedures
REVIEW the results	How do we measure how we have done?	**i-Actions (i-A)**	Specific forms or templates
REMEMBER & learn to improve	How can we do better?	**i-KnowHow (i-KH)**	Knowledge items to look for

Links: AAGC tales will often conclude with a hyperlink to a YouTube video, an external web site or a sample of how the content might appear in our QW sample architectural project.

But enough introductions! Let's get started with Tale #1.

Making Commitments
Assurance Letters

"The evidence before this Commission is clear that the waterproofing of the roof failed virtually from the outset. The logical course of action would have been to undertake early and effective remedial measures to protect a valuable asset. Successive owners neglected to do so, and the consequences of that neglect were tragic."[1]
- The Honourable Paul R. Belanger, Commissioner, Report of the Elliot Lake Commission of Inquiry, 2014

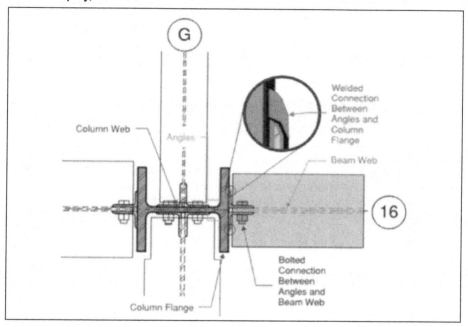

Figure 14 - Location of the Elliot Lake Collapse[2]

"Their report attributes the cause of failure to an undersized steel beam in a Cantilever-Suspended Span arrangement and inadequate buckling resistance of the beam compression flange. In addition to the technical errors were a number of procedural deficiencies in the project delivery system."[3]– Station Square Commission Report, 1988

---

1   Hon. Paul R. Belanger, Report of the Elliot Lake Commission of Inquiry, Queen's Printer for Ontario, 2014, Section II, Causes of the Collapse, http://www.elliotlakeinquiry.ca/report/Vol1_E/ELI_Vol1_Ch03_E.pdf
2   Hon. Paul R. Belanger, Vol. 1, p. 38
3   Wilson, James, N., "Supermarket Roof Collapse – Burnaby, BC, Canada, in Wikipedia at http://bit.ly/ZQu6Kf

Figure 15 - Station Square Roof Collapse, 1988[4]

"The Commission of Inquiry studied the responsibilities and obligations of the different members involved in the design and construction for the Save-on-Foods store. A trend that brought concern at the time of construction, which is one that continues today, is the fragmentation of practices. With a construction industry that involves many different contractors, coordination between all of them can bring great challenges. Also, varying organization models in the construction industry make it difficult to delegate responsibility and liability."- Station Square Commission Report, 1988[5]

What's the point? I had not initially intended to have a "tale" about professional assurance letters as they are not yet widespread and I have no dramatic personal experience of them. But as I was completing this manuscript one of my beta readers reminded me that the first report cited above had just been published. It analyzes the collapse of a 33-year old shopping mall roof in central Ontario in 2012 that killed two persons and injured several others.

Twenty-four years before this roof collapse in Elliot Lake, a shopping mall roof also collapsed thousands of miles away, in British Columbia (BC) on opening day . There were no fatalities but 21 people were injured. The similarities do not end there. In both projects, rooftop parking collapsed. In both projects there was a dramatic absence of coordination between the various architects, engineers and clients. At Elliot Lake the lack of coordination manifested itself throughout the life of the structure. Even though the Elliot Lake building was built about nine years before the Station Square structure collapse, issues that started during construction eventually brought it down.

The Station Square collapse in 1988 and subsequent inquiry caused the architectural and engineering professions in BC to undertake a major overhaul of professional responsibilities, especially the coordination of those responsibilities. The government of BC introduced a series of professional assurance letters that have been copied by some other Canadian jurisdictions, but not Ontario, although that is a recommendation of the Elliot Lake inquiry.

---

4   Wilson, James N., Ibid.
5   Wilson, James N., Ibid

In the years leading up to the Station Square event, and the 33 years from construction to the deadly collapse of the Algo Centre Mall in Elliot Lake, there has been unrelenting economic pressure on architects and engineers to reduce their fees and scopes of work. Each professional responds to this in her/his own way, and the large majority have maintained reasonable professional standards in the public interest. The nineteenth century concept of the "master builder" has been continually eroded and the delivery of buildings in many places has become increasingly diffuse and uncoordinated, culminating in these tragedies and a handful of others.

What are the principles & best practices? This is very serious stuff. My analysis may seem simplistic, but there is lots of detailed discussion and recommendation in the reports cited, and from many other sources.

In response to the Station Square building collapse in BC, government, in consultation with the design and construction professions, developed a trio of assurance letters, lettered A, B and C for simplicity, that are now required for all but the smallest buildings in the province. The province of Alberta has created clones of the BC forms. For simplicity we will use the BC forms for discussion.

Schedule A confirms the project's owner has retained a qualified professional called the "Coordinating Registered Professional" (CRP) to coordinate the services of the consultant team for both the design and field review of the project. The CRP countersigns the Schedule A to confirm understanding of the scope and undertaking.

SCHEDULE A

Forming Part of Sentence 2.2.7.2.(1), Div. C of the
British Columbia Building Code

Building Permit No.
(for *authority having jurisdiction's use*)

CONFIRMATION OF COMMITMENT BY OWNER
AND COORDINATING REGISTERED PROFESSIONAL

Notes: (i) This letter must be submitted before issuance of a *building* permit.
(ii) This letter is endorsed by: Architectural Institute of B.C., Association of Professional Engineers and Geoscientists of B.C., Building Officials' Association of B.C., and Union of B.C. Municipalities.
(iii) In this letter the words in italics have the same meaning as in the British Columbia Building Code.

Re: Design and *Field Review* of Construction
by a *Coordinating Registered Professional*

Figure 16 - BC Schedule A top of Page 1

Notice that the commitments in Schedule A are jointly undertaken by the Owner and a Coordinating Registered Professional (CRP), usually an architect. This joint undertaking is specifically designed to eliminate "I thought you were…", "But you said you were…" kinds of discussions.

As to what the Owner and CRP are agreeing to:

The undersigned has retained _____ as a *coordinating registered professional* to coordinate the design work and *field reviews* of the *registered professionals of record* required[1] for this project. The *coordinating registered professional* shall coordinate the design work and *field reviews* of the *registered professionals of record* required for the project in order to ascertain that the design will substantially comply with the B.C. Building Code and other applicable enactments respecting safety and that the construction of the project will substantially comply with the B.C. Building Code and other applicable enactments respecting safety, not including the construction safety aspects.

Figure 17 - BC Schedule A Scope

The definition of CRP sounds remarkably like the traditional role of the architect as prime consultant!

Essentially, the CRP is committed to coordinate the entire design team's design work as well as their field review, with the expectation of substantial compliance with the building code.

Just above the joint signature lines for the Owner and CRP is an interesting paragraph:

The *owner* and the *coordinating registered professional* understand that where the *coordinating registered professional* or a *registered professional of record* ceases to be retained at any time during construction, work on the above project will cease until such time as
  (a) a new *coordinating registered professional* or *registered professional of record*, as the case may be, is retained, and
  (b) a new letter in the form set out in Schedule A or in the form set out in Schedules B, as the case
   may be, is filed with the *authority having jurisdiction*.

Figure 18 - BC Schedule A Continuing Commitment

This inclusion provides for the continuous presence of a CRP on the project – if the CRP leaves or is relieved, everything stops until a new CRP has been appointed and has in every sense taken over the project from the predecessor CRP (which can be tricky – think about taking over someone else's design part way through construction.)

Only once have I left a project where I was the CRP. The details are unimportant except to say that I had lost faith that I could meet the commitments of Schedule A due to the actions of the Owner. I was obligated to go to city hall and withdraw my Schedule A (and others mentioned below), which resulted in an immediate "stop work" order until I was replaced. Nasty indeed.

So we have a CRP in place, but no design team. Nothing in Schedule A refers specifically to the architect's technical work scope; none of the other consultants is covered as yet. That's what Schedule B's are for.

Schedule B requires each consultant on the project, including the architect, to clearly define the scope of their work using standard terms, and to commit that their design of that scope will comply with the building code and that they will provide the field review necessary to determine construction complies with the building code.

SCHEDULE B

Forming Part of Subsection 2.2.7, Div. C of the
British Columbia Building Code

Building Permit No.
(for *authority having jurisdiction's* use)

**ASSURANCE OF PROFESSIONAL DESIGN AND
COMMITMENT FOR FIELD REVIEW**

Notes: (i) This letter must be submitted prior to the commencement of construction activities of the components identified
    below. A separate letter must be submitted by each *registered professional of record*.
  (ii) This letter is endorsed by: Architectural Institute of B.C., Association of Professional Engineers and
    Geoscientists of B.C., Building Officials' Association of B.C., and Union of B.C. Municipalities.
  (iii) In this letter the words in italics have the same meaning as in the British Columbia Building Code.

Figure 19 - BC Schedule B

For convenience, one Schedule B is used for all of the Consultants – they indicate their area(s) of responsibility by initialing against defined disciplines and applying their professional seal:

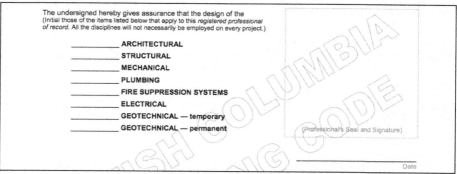

The undersigned hereby gives assurance that the design of the (Initial those of the items listed below that apply to this *registered professional of record*. All the disciplines will not necessarily be employed on every project.)

_____ ARCHITECTURAL
_____ STRUCTURAL
_____ MECHANICAL
_____ PLUMBING
_____ FIRE SUPPRESSION SYSTEMS
_____ ELECTRICAL
_____ GEOTECHNICAL — temporary
_____ GEOTECHNICAL — permanent

(Professional's Seal and Signature)

_____
Date

Figure 20 – BC Schedule B Discipline Identification

Each professional of record makes the same commitments in relation to their discipline as does the CRP:

components of the plans and supporting documents prepared by this *registered professional of record* in support of the application for the *building* permit as outlined below substantially comply with the B.C. Building Code and other applicable enactments respecting safety except for construction safety aspects.

The undersigned hereby undertakes to be responsible for *field reviews* of the above referenced components during construction, as indicated on the "SUMMARY OF DESIGN AND FIELD REVIEW REQUIREMENTS" below.

Figure 21 – BC Schedule B Commitments

Each professional of record has areas of responsibility identified in the building code and either initials against the entire list, or strikes out any services by others. For example, in the architectural list below, the architect professional of record might strike through 1.6, structural capacity of architectural components, in cases where others engineer guardrails, handrails, etc.. In such cases, the building authority would then expect to see a Schedule B from that professional of record, with everything struck through except 1.6.[6]

One silly discussion can occur when, for example, the architect strikes through 1.11 Elevating devices because there is no elevator in the building. True enough, but it's like a double negative – if there are none there then you don't have to strike it out, because there are none there!

---

6   In British Columbia there exists a Schedule "S" (for "Specialist" Consultant) that may be used instead of a Schedule B.

SUMMARY OF DESIGN AND FIELD REVIEW REQUIREMENTS

(Initial applicable discipline below and cross out and initial only those items not applicable to the project.)

_____ **ARCHITECTURAL**
1.1 Fire resisting assemblies
1.2 *Fire separations* and their continuity
1.3 *Closures*, including tightness and operation
1.4 Egress systems, including *access to exit* within *suites* and *floor areas*
1.5 Performance and physical safety features (guardrails, handrails, etc.)
1.6 Structural capacity of architectural components, including anchorage and seismic restraint
1.7 Sound control
1.8 Landscaping, screening and site grading
1.9 Provisions for fire fighting access
1.10 *Access* requirements for *persons with disabilities*
1.11 Elevating devices
1.12 Functional testing of architecturally related fire emergency systems and devices
1.13 Development Permit and conditions therein
1.14 Interior signage, including acceptable materials, dimensions and locations
1.15 Review of all applicable shop drawings
1.16 Interior and exterior finishes
1.17 Dampproofing and/or waterproofing of walls and slabs below *grade*
1.18 Roofing and flashings
1.19 Wall cladding systems
1.20 Condensation control and cavity ventilation
1.21 Exterior glazing
1.22 Integration of building envelope components
1.23 Environmental separation requirements (Part 5)
1.24 Building Envelope, Part 10/ASHRAE Requirements

(Professional's Seal and Signature)

_____
Date

Figure 22 – BC Schedule B Architect Scope

Schedules A and B are submitted at the building permit stage; no building permit will be issued until they are delivered.

Schedule C's, which are prepared at the completion of the project, are discussed in Tale #66, called "Completing Commitments."

RECORD the journey	Assurance letters beginning & end	Summary Principles & Tools	
RESOLVE the issues	Letter items stuck through	i-WorkPlan (i-WP)	Procedures for assurance letters
REVIEW the results	"C" letters assure commitments have been met	i-Actions (i-A)	Transmittal of assurance letters
REMEMBER & learn to improve		i-KnowHow (i-KH)	Local interpretations

Links: BC Assurance letters are at http://bit.ly/aagc-BC; Alberta building code schedules are at http://bit.ly/aagc-alta;

## 02

Organizing
Assess & Manage Risk

"It's a bit late in the design for you to tell me you have never designed a swimming pool!" – Principle to Project Architect

Construction is a risky business in which architects cede most control to contractors. But they can control design risk by being clear about it internally and ensuring risks identified during design are resolved by the final design, or monitored closely during construction. In a design team environment it may not always be clear what risks are inherent in the site, the design, the client, etc.

Your i-WorkPlan should include from as early as possible (i.e., schematic design) a log identifying project risks as they occur. Items can be added and closed as they are resolved:

Project Risks

#	Item	Solution	Remarks
1	Stream adjacent site subject to spring floods	Landscape architect to include mitigation. Site plan may need to include extra setback	Consult Fisheries and Oceans Canada early re setback and rip rap design
2	Older structure to the west	Structural engineer & geotech to review re best approach to temporary support. Specify contractor to measure & monitor settlements & cracking.	Advise client that there may be costs associated with repair to cracks of neighbor building.
3	Want to use wall cladding material new to this office	Research: litigation; best practices. Review detailing with building envelope consultant.	
4			
5			
6			
7			
8			
9			
10			

Figure 23 - Being Clear about Risks - A Project Risk i-A

In the example above, many of the items can be assigned to the relevant consultants for resolution. Additional challenges can be added as they arise. Of course, this requires that they be resolved, not forgotten! Alternatively, each individual risk item can become its own i-A.

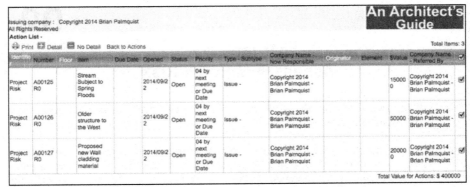

Figure 24 - Project Risks as i-A's - Note $ Values

Contractors also identify risks, first during estimating (estimator), then during contract finalization (project manager) and finally during construction (superintendent). When each Risk is considered as its own i-A, they can be individually assigned and a $ value can be attached to them. As each is resolved, it can be closed and it and its $ value will drop from day-to-day view.

RECORD the journey	Using risk management procedure in i-WP	Summary Principles & Tools	
RESOLVE the issues	An i-A for each risk until resolved	i-WorkPlan (i-WP)	Specific procedures
REVIEW the results	Align risk patterns with 10C's	i-Actions (i-A)	i-A risk template
REMEMBER & learn to improve	Refine i-WP re risk experience	i-KnowHow (i-KH)	i-KH for emegent risks

Links: To access the form "Risk Review", go to http://bit.ly/1i69VuV . Click on the blue http file name at the bottom left of the screen.

## 03

Organizing
What do we Build from? Not the Specs!

"The specifications just sit on a shelf in my office, I never look at them. Far as I am concerned, if it's not on the drawings it's not in my scope." – Contractor to Architect

"Read the words, then do what the words say!" – Architect to Superintendent

This fundamental misunderstanding of the essential role of specifications is unforgivable but also understandable. Many consultants do not write specifications, write them badly or place them on their drawings – either way the contractor is encouraged to ignore separate written volumes.

In the simplest terms: drawings show where and how many; specifications show what and how. Both types of information are essential to properly construct.

This misunderstanding is exacerbated by the fact that many consultants do not know how to write specifications – if they do them at all, they outsource them or assign them to one in-house "specialist." Neither is a good idea in terms of mentoring the next generation.

Outsourced specifications tend to reference every standard known (e.g., CSA, ASTM, ANSI, CGSB, etc.) in the clear belief that all of these standards are both in the architect's office and on the construction site, which is not a practical possibility. To complement the raft of standards, these specifications include statements like *"... where there is a conflict between two standards, the more onerous will apply."* Really? Give me a break! It is hard not to believe that a statement like this means that either the consultant does not know which standard is more onerous, or does not care. The specifications may also complete the picture with statements like *"install per manufacturer's instructions"*....and the instructions are where?

Without in any way diminishing the value of experience, the alternative to outsourcing specifications, the in-house expert, usually prevents younger architects in the field from learning how the words relate to what happens in the field.

It is probably clear from my words above that I favor having specifications broadly written by the consulting firm doing the design. The best approach is to have the project architect write the specifications, then have them reviewed by a more senior peer. I also favor minimizing the number of referenced standards to those the firm actually knows, has to hand and understands. And instead of "install per manufacturer's instructions", I favor incorporating the actual instructions in the specification. All of this can be accomplished with master documents or i-Actions.

When first challenged to write my own specifications, I started by buying a "master specification." There was nothing much in Canada at the time, so I purchased an American product, including a specifications processing program called "Masterworks," which still exists. There were two great things about this purchase:

Firstly, because the specs were American, I had to substantially rewrite them for Canada. This caused me to review their content and start to think carefully about what the words said and whether they were useful in the field. I deleted tons of verbiage and substituted a smaller volume of meaningful content. Because I know the words I have written, and see how others interpret them in the field, and because every spec section is a "master", I regularly tweak the words as I see what was understood or not.

Secondly, the specifications processing program automated spec production so well that I could produce a professional looking, indexed 200-page spec in two working days. While I generally favor web-based products, Masterworks' (and similar products') manipulation of complex content on a MS Word base pays for itself on your first spec. I also had a very profitable side business for some years writing specifications for architects who had never learned themselves. $5,000 for two days hard work pencils out nicely.

If we want to mentor younger architects we need to reinforce the connection between specifications and what happens in the field. Specifications that make sense and are detailed will assist that effort. As an example, here is a short excerpt from my "master" rough carpentry specification, which I have set up to be able to be edited based on the project specifics. Every clause in this sample was written by me based on field observation:

D.	Lumber plates:
1.	Use preservative treated sill plates, set atop continuous 1/4""polyethylene sill gasket, density 1.1 as manu. by ProPack, Delta (946-9116). If concrete is uneven, use multiple layers of gasket to achieve air seal. Seal inside wood-to-concrete junction continuously with Chem Calk 900 or other preapproved sealant.
2.	Where using a pressure treated (preservative treated) sill plate, the moisture content (m.c.) of the plate may be as high as 25% so long as the m.c. of ALL abutting wood, including sheathing, is as much below 20% as the plate is above 20%. For example, a 25% preservative treated plate is acceptable where the balance of adjoining lumber is ≤15%.
3.	To protect sill plates from taking up water from concrete topping, install a continuous piece of plate polyethylene (typ. 16"" wide) to cover the inside face of the plates & the partial perimeter of the floor sheathing just before topping is poured.
4.	To keep wood-frame walls drier and facilitate air/vapour barrier continuity, drape the top plate of each wall with plate polyethylene.

Figure 25 - Specification Excerpt - Every Clause Based on Field Experience

There is no magic to this approach. Every clause is based on experience with what has and has not worked. A bidder might disagree with what I have specified but it is clear and easy to price. A superintendent will easily be able to follow these instructions.

RECORD the journey	Write specs that you understand	**Summary Principles & Tools**	
RESOLVE the issues		**i-WorkPlan (i-WP)**	Specific procedures
REVIEW the results	Compare field experience to written specs	**i-Actions (i-A)**	Could be used to assemble specs
REMEMBER & learn to improve	Refine specs based on project experience	**i-KnowHow (i-KH)**	

Links: The latest generation of the spec writing software I used is available at http://bit.ly/aagc03-1 .

## 04

Organizing
It's All About Time

"Every month I feel like I've left a bunch of fees on the table, but I can never seem to capture them!" – One poor architect to another (over beer)

What's the Point? I mention in a later tale about charging for your value engineering efforts and other additional services – I have been successfully doing so for years. If you are not already keeping track of your time down to the tenth of an hour, start now! Lawyers have been doing it for years, and charging for it for years.

What are the principles & best practices? There are many cheap apps that will track time. Regardless of whether your time is currently tracked carefully by you, your employer, your clients, whoever, start tracking it right now, and to the tenth hour. Some useful things to consider in choosing software:

Portable – Much of your time is spent in the field and other offices. And let's be frank – you probably bring home some homework. Make sure your tablet and smartphone can run the app. I do my time tracking on my tablet because that's what I bring into the field, and I always bring it home. I haven't tracked time on a laptop for 7 years.

Two phase tracking – By this I mean the ability to track first, a project, secondly, a phase of the project. If one of those phases is "additional services", and you can create a report by project and phase, you will have a concise additional services report for each project whenever you need it.

Notes – I hate duplication of effort, so keep detailed notes of meetings on my tablet. But I do not consider a one-line time tracking summary note as duplication. This becomes a printout to the time sheet that may be attached unedited to the monthly invoice.

Integrated – Ideally your mobile app should synchronize with fully featured accounting software, BUT, if you are solo or a small firm, think hard before you pay for that integration. If you pay $4.99 for an app that tracks your eight projects really well and from which you are able to create invoices in 1 hour each month, is it worthwhile to spend $1,000 to reduce that time by half, after learning to use the new software? At a certain size of firm such integration may be desirable, just think carefully what that point is. In my experience, many sophisticated systems exist for the benefit of office staff over the folks who earn the money.

Rounded (Up) – Many apps have simple settings that will round up so that a 34-minute meeting becomes 0.6 hours (in my tenth hour approach). Clients are used to tenth hour invoices from their lawyers, so a bill for 6.7 hours will not look strange.

Backed up – People lose time records all the time. Web-based apps may backup to the cloud, but you will achieve the same result by emailing to-day's time sheet at the end of each day; the most you will have to recreate from memory is one day of data. I usually do mine on the bus on the way home. Professional services gurus have recommended daily time sheets for years for best monitoring of work flow and capture of value – hard to argue against.

History - Time tracking will quickly inform you where you are efficient and where not. It will inform all of your fee calculations and give you a database that will assist you when negotiating.

Most time tracking apps allow you to quickly switch projects, so when you are interrupted by a call from Project A while you are doing design drawings for Project B, you can pause B and capture A. These apps understand how Consultants (read lawyers) work, so can usually print out a report per project, to summarize your efforts and segregate your costs.

When I was in private practice and using tenth hour time tracking religiously, I personally billed at least $1,000.00 per month in additional services that were only captured by that diligent time keeping, which quickly becomes second nature when you realize its value.

Clients in particular love to ignore time. When faced with a negotiation in which you are being pressed to reduce fees, if you have a fee calculation that is based upon true historic time tracking, then you are more able to say things like *"...the model is budgeted at $3,000 based on my experience, including $500 of my time coordinating; perhaps you should look after that to save the $500."* - or - *" ...the contractor advises he wants us to attend weekly site meetings, we could save you $x if we attended every 2nd week, at least until the project is further along."* Of course, you should not propose changes to scope that cut into basic services or services mandated by your profession.

RECORD the journey	Record all the time all the time.	**Summary Principles & Tools**
RESOLVE the issues	Charge for your additional services	**i-WorkPlan (i-WP)**
REVIEW the results	Identify your real time cost of different work scopes	**i-Actions (i-A)**
REMEMBER & learn to improve	Ditto	**i-KnowHow (i-KH)**

Links: My personal choice for mobile timing is QuickTimer ($4.99) at http://bit.ly/aagc05-1 - there are many others.

## 05

Starting
What Contract is this Anyway?

"It's a very straightforward project, but unfortunately we used most of the construction phase fees during design, so you only have a half day a week to manage the construction phase." – Principal to Intern Architect

"Read the words, then do what the words say!" – Contractor executive to project team

What's the Point? Seriously? Do we have to talk about contracts? Really?

The sad reality is that with amazing regularity those of us expected to administer a construction contract efficiently (read cost effectively) have little or no idea what we are administering, nor what the professional services contract says we are supposed to do as part of construction phase services.

As an occasional teacher of intern Architects and students, I used to jump right into meaty stuff like claims for payment, changes to the work, submittals, etc. (all coming later in this e-book). Noticing some blank stares, I began to ask who in the audience had read the contract they were administering – always less than half. Further questioning revealed that in many cases principals of firms did not feel it was important that their interns knew what they were administering!

What are the principles & best practices? The world would be a simpler place if the contractual relations between architects and their clients, then between clients and contractors, were standard. They used to be, and many architects and teachers of architects act as if that era still exists in a vacuum called "Design/Bid/Build" or "Stipulated Sum." That's the simple linear approach where the client prepares a program, the consultants prepare and document a design that expresses the program as built form, several contractors indicate how much money they want to build the building, one is hired and builds it.

But reality is increasingly a bit more complex. Before we get into the other complexities, there is some good news. There are still really only four parties to design and construction: the client; the prime consultant; the contractor; and the public.

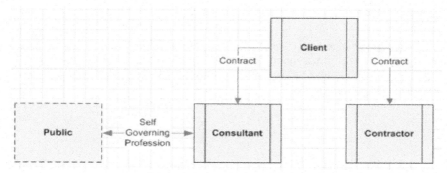

Figure 26 - Four Parties to Design & Construction

The client may be an individual (think homeowner), a group of individuals (think co-op), or a company, nonprofit society, etc. Regardless of the client-side structure, the designer and builder need to identify who the owner's representative is – the individual who is the filter between the client and the designers and builders.

The owner's representative, a.k.a. client representative, is the individual who is empowered to make all of the decisions that a client needs to make during design and construction. He/she may not personally sign the cheques for design and construction, the change orders, etc., but will be the person who recommends payment and approves changes.

Sounds simple, but seldom is. In a home renovation for a family, is it the husband or wife? In a non-corporate group endeavor, who is the spokesperson? In a corporation, who speaks for the corporate entity. Note that even when you have identified the owner's representative, she/he may change at any moment. If this occurs or appears to occur, ask again who is the owner's representative. Where a client insists on more than one representative (think school board), you will need to copy every communication to every person purporting to be a client representative, even if they say "no need."

A major target reader of this e-Book is the second leg of the pyramid, the prime consultant. When I started my career the architect was almost always the prime consultant, managing a group of subconsultants generally including structural, mechanical/plumbing and electrical engineers, a landscape architect and occasionally specialist consultants like an acoustician in the case of an auditorium-type space.

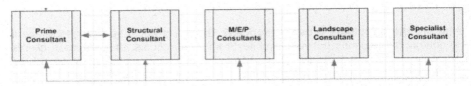

Figure 27 - Traditional Prime/Sub Constultants

In this historic form of relationship, the subconsultants have contracts with the prime consultant for their specialized services. The prime consultant traditionally earned an additional fee as the manager of the subconsulting group. This fell somewhat into disfavor for two main reasons: firstly, some clients felt they could save fees by hiring specialist consultants directly; secondly, a few architects delayed passing payments through to subconsultants, in a few cases failed to pay them entirely (and were eventually disciplined). So subconsultants began to "report to the architect" but "bill the client."

The contractual arrangement does not change the responsibilities. Imagine a building design where none of the engineering services is coordinated with the architectural design or with each other (at this point, some contractors will say that's exactly what happens all too often!).

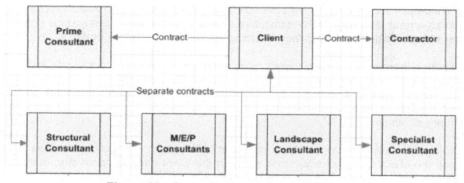

Figure 28 - Consultants retained by Clients

The insurers of consultants have exacerbated this situation by pointing out that where the prime consultant retains the specialist consultants, the client can only sue a specialist consultant through the prime consultant. True, but as a prominent construction lawyer in my city once advised me, *"Regardless of the contract, if you perform your services well, you will be fine. And if you mess up, you will be in trouble regardless of the contracts."*

The reality is that there is little if any savings achieved by hiving off the specialist consultant contracts from the prime's contract. Nonetheless, it may occur. Don't fight it, but don't reduce your coordination fee either.

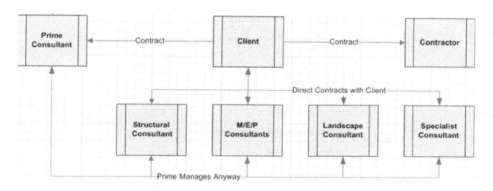

Figure 29 - Client Hires Subconsultants but Prime Directs them Anyway

Occasionally a consultant retained directly by the Client then decides not to be managed by the prime consultant, i.e., not to reveal scope of services, attend meetings, provide field review, etc. Most professional associations' codes of conduct require that a consultant retained on a project with a prime consultant disclose and fulfill scope of services to the prime or managing consultant, but need not disclose fees. Nothing wrong with that, I don't need to know what her/his fees are. I have received the occasional consultant work scope so heavily redacted (i.e., black felt penned) in relation to fees that I feel like a spy for CSIS/CIA/MI5. As long as I can see what she/he is up to, the "commercial terms" (as contractors like to call the money) are up to them – except, of course, if she/he is not providing the minimum scope of services needed for their area of expertise. I have on several occasions obtained scope and fee increases for subconsultants simply by saying *"This does not meet professional standards. Either*

39

you increase your scope, or I will have no choice but to report you to your association."
Works a treat, and usually the consultant thanks you later and quietly, since she/he
has obtained significant extra fees for doing the job they should be doing.

Although there may be many variations in agreements between consultants and clients, they should have these principles in common: Consultants design to the client's
program of requirements and to the established budget as well as applicable codes
and regulations. Where program and budget are out of sync, the consultant needs the
client's direction.

During the course of construction, consultants: monitor construction activities and
keep the client informed; adjudicate between the client and the contractor; manage
confirmation of design through vehicles such as submittals; review and recommend
proposed changes to the work; periodically review the unfolding construction; respond
to reasonable information requests from the contractor or client; review the contractor's bills and recommend how much the client should pay; assist both the client and
contractor to complete the construction and gain occupancy; determine when the
project is substantially complete/performed, i.e., is "ready for the use for which it was
intended". They also have an additional duty to protect the public interest, usually
expressed in the form of zoning and building regulations and adherence to their embedded codes and standards.

In a typical project, there are up to 100 discrete activities that an architect/prime consultant will be involved in – and that's just during the course of construction!

Process Name	Procedures
RECORD Construction contract administration services	9
RECORD Consultant construction phase services	7
REMEMBER live knowledge, lessons learned and best practices	2
REORD Course of construction quality management	1
RECORD services - changes to the work	2
REVIEW Changes to construction	1
REVIEW Changes to the Work affecting Quality	1
REVIEW Requests for Change	2
REVIEW proposed changes with regulatory agencies	2
REVIEW material samples	1
Consultant Submittal REVIEW	2
REVIEW Critical Products	1
REVIEW proposed substitutions	1
REVIEW & Respond to Requests for Information	1
RECORD Site Instruction (SI)	1
RECORD Contemplated Change Orders (CCO)	1
RECORD Change Order (CO)	1
RECORD Change Directives (CD)	1
REVIEW Contractor/Owner responses to CD	1
REVIEW Contractor schedule of values	4

**Figure 30 - Partial List of Architect's Construction Administration
Considerations**

The third leg of the design and construction structure is the contractor. Many contractors, especially at the site level, forget that consultants do NOT work for them. And sometimes they deliberately forget, as when they ask for an unreasonable level of services (read detailed inspections of every room at several stages, de facto punchlists near completion, etc.).

Figure 31 - There is usally no Contract Between the Consultant and Contractor

This situation is further complicated by a basic tenet of contracts between clients and contractors. Having worked diligently for the client during the design phases of a project, the consultants are expected to become "independent arbiters" of the contract between the client and contractor as soon as construction starts. This inevitably means that they have to periodically tell the client (who originally hired them and is still paying them during the construction phase) that they are wrong and the contractor is right. And they even have to do that when the client's comeback could well be "Why didn't you cover that off in your design?" Conversely, they have to periodically tell the contractor when a request for services or claim is unreasonable in relation to the project's contracts.

There are many variations in construction contracts, but they have these concepts in common:

The contractor: provides a price to construct the project, either fixed or budgeted, and updates that pricing as the project unfolds; is in control of the construction site and all activities on it, including safety and security; determines and executes the "means and methods" of construction needed to execute the design as documented, including selection of suppliers and installers and their specific installation procedures; contracts with the various suppliers and installers, and pays them or manages their accounts in the case of arrangements such as construction management; is responsible for ensuring that the construction completely conforms to the design (and applicable codes, standards, etc.); is responsible for obtaining substantial completion/performance of the work (but is usually not responsible for getting occupancy approval); and is responsible to correct defects that arise during a defined warranty period (usually one year from the date of substantial performance).

Process Name	Procedures
RECORD Construction contract administration services	9
REMEMBER live knowledge, lessons learned and best practices	2
REORD Course of construction quality management	1
RECORD Construction management QA Requirements	4
RECORD services - changes to the work	2
REVIEW Changes to construction	1
RECORD Requests for Change (RFC)	1
RECORD submittals list	1
RECORD Contractor submittal schedule + submittals	2
RECORD Contractor Prepared submittals	2
REVIEW Critical Products	1
RESOLVE contract amount or time from submittal	1
REVIEW Site Instructions (SI)	1
REVIEW Contemplated Change Order (CCO)	1
RESOLVE Issues with Preventative Action Reporting	1
REVIEW Contractor schedule of values	4
REVIEW claims for payment	1
REVIEW moisture management strategy	1

Figure 32 - Partial List of Construction Admin. Processes for Contractors

The client in a construction contract: takes care of insuring its own interests, such as ongoing operations in a building under renovation; pays the cost of construction as/when recommended by the consultants or others acting as payment certifiers; responds to reasonable information requests from the contractor or consultants; and acts as the owner when required by other parties such as the public, municipal and other levels of government.

Some other contractual arrangements include:

Construction Manager (CM): The concept of the construction manager has been around for decades, but really took hold in the days of high inflation in the 70's and 80's, when clients became convinced they could save time and money by having professional management of the construction contracts separate from that traditionally provided by consultants. This thinking gave rise to two basic construction management variations:

Construction Manager for Services: Also known (incorrectly) as "CM as Agent", in this arrangement, the CM uses expertise to develop and manage budgets, vet potential trade contractors and negotiate or tender contract scopes and prices with them. Some consultants act as CM for services – they do not have direct contracts with the trade contractors doing the work, rather these rest with the client:

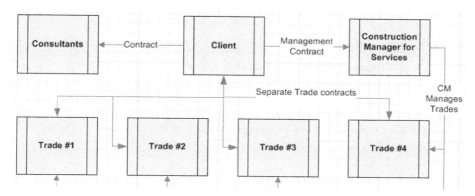

Figure 33 - CM for Services (as agents)

I have been involved in CM for services projects. One highly visible difference as compared with traditional design/bid/build arrangements is the monthly claim for payment. Whereas the conventional claim from the contractor is seldom more than a dozen or so pages, a CM for services claim will include detailed invoicing from every trade contractor and major supplier, of whom there will likely be dozens. We're talking one or more 4-inch (100mm) binders full of backup!

Construction Manager for Services and Construction: Also known (incorrectly) as "CM at risk", but old habits (and phrases) die hard. As far as what it means, some clients get nervous, either about having all those contracts with all those trades and suppliers, or about having a budget rather than a fixed price. So it sometimes happens that as part of the CM decision, or sometime thereafter, the client will ask for the budgeting, scoping, tendering and negotiating work to be converted into a fixed price. This is the "Risk" and it has the effect of reducing the pool of potential CM's, as many consultants will choose not to take on construction risk, or are unable to obtain the necessary bonding and/or insurance.

In addition to these more common variations, there are several others:

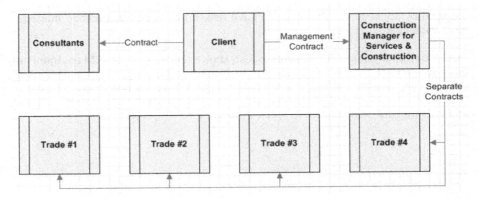

Figure 34 - CM for Services & Construction (at Risk)

Increasingly, complex buildings such as hospitals are being procured through arrangements where the contractor retains the Architects and Engineers (A/E above). Since the A/E team's communications with the client are reduced, sometimes the client retains a separate consultant team to monitor the design and construction.

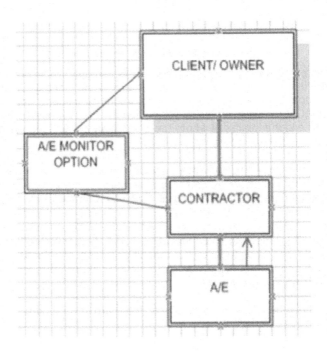

Figure 35 - Design/Build + Monitor

Very Occasionally an Architect or Engineer is so versed in a certain type of design and construction as to be the obvious choice to retain the construction Contract – Design-Build with the Designer on top (literally). I am aware of two such successful ventures in my local marketplace – one involves the specialized needs of ski resorts, the other the particular requirements of first nations clients.

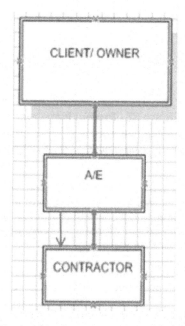

Figure 36 - Consultant Run Project

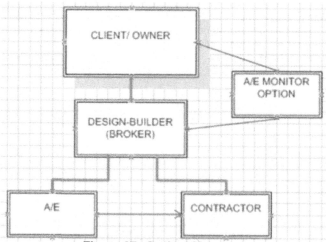

Figure 37 - Project Manager

I find this arrangement challenging, but it persists, more often in public sector facilities where the client's knowledge may be limited (e.g., a small community developing a big arena).

When you are working on a project that has a person or company, often called a Design-Build Broker or Project Manager (sad misuse of the professional term), who stands in the midst of all communications and decision-making without seeming to contribute much, this is what you are probably working with. Be prepared for much duplicated and side tracked communications.

Conversely, at its best this arrangement involves a "Design-Build Broker" whose deep expertise obtains for the client an optimized project at minimum cost.

Joint ventures are not too popular at the moment, but you may stumble into one. In a true joint venture the consultants and contractor(s) make a joint proposal to design and build a facility, usually a complex building like an arena or hospital. They form a company for the specific purpose of this single project, and although the JV may appear to evaporate after the project is completed, the fact is that the participants remain liable for their actions for the duration of the project's life, except in the case of a statute of limitations.

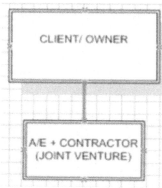

Figure 38 - Joint Venture

Where a JV's involvement extends to financing and/or operating a facility for a specified time period for a specified annual cost, the JV has "morphed" to become a P3 – Public Private Partnership.

Some contractors have in-house professionals to design their work. Many single-family home tract developers operate in this fashion, especially as they may be involved in repetitive designs. When such an organization takes on a project outside their comfort zone, they may engage an Architect or other consultants to flesh out their expertise. This can be an enjoyable experience where the in-house professionals act as communication intermediaries between the builder and the specialist architect.

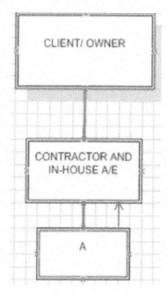

Figure 39 - In-house Consultant

I am sure there are many other variations on the theme, but what you want to be doing is figuring out where in this continuum your current project fits – what are the relationships between the client, consultants, contractor and the public.

Speaking of the public, you may have noticed the only links to the public are through the Consultants:

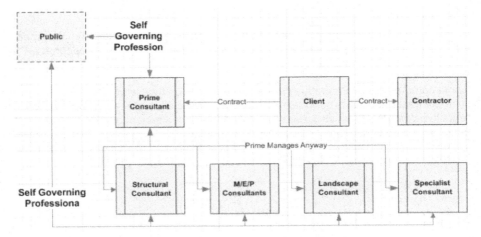

Figure 40 - The Self-Governing Professions

Regardless whether any state, provincial or federal jurisdiction has granted the design professions self-governing status, the expectation is that professionals will in all their work uphold the interests of the general public. This most often means ensuring that executed designs conform to the various local, state/provincial and federal regulations of the project's location. The key concept here is that this obligation does not necessarily extend to clients and contractors. Of course, most clients and contractors choose to operate in a fashion that upholds the public interest, but they are not obligated to in the same way that registered professionals are.

If this e-Book sells well, my next effort may be an architect's guide to business, which will expand significantly on this information. Here I am just trying to provide enough context so that less experienced readers are not overwhelmed by the contractual context they work in.

RECORD the journey	Ensure the i-WP matches the forms of contract	**Summary Tools**	**Principles &**
RESOLVE the issues	Remember contractual relationships	**i-WorkPlan (i-WP)**	Ensure i-WP templates are a match for the contract forms

## 06

Starting
Value Engineering / Design Assist

"I could save you lots of money if the architect wasn't so uptight about the VOC's – whatever they are" – Contractor to Client

The Contractor had tabled his latest proposals to save the client money on our project. As the building envelope consultant, I was mainly a spectator to the tussles between the contractor and the beleaguered designer. Most of the changes proposed as money-saving involved major redesign by the consultant team, plus revisiting the planning permits, which were premised on specified massing, materials and colors, what architects and planners like to call "form of development." This latest proposal was going to lose a precious LEED® credit.

What's the Point? Throughout the course of most projects, team members will have ideas about how to improve the value of the project. Some will involve cost or time savings; others may involve extra costs or schedule time in exchange for a higher quality product. Some ideas will arise during the normal course of design development, and will be considered at regular design coordination meetings. But other ideas may arise out of sequence, i.e., after documents have been issued for bids or for construction.

Value engineering is the most common phrase used to describe this out of sequence process. Contractors are often nervous about that phrase because it suggests involvement in engineering, hence design, hence design liability that most contractors are not insured for. Value engineering also has a bad name, equated with chopping and hacking at a design to pursue cost reductions. "Design assist" seems to avoid this stigma. In the hands of a more experienced contractor and client, design assist can be down right pleasant! On the other hand, some may manage it as brutally as the worst examples of value engineering.

Some years ago, the government of my province addressed rapidly escalating school construction costs by dramatically cutting back on design standards using value engineering processes, including even the expected lifetime of new schools. In the face of independent standards that suggested schools should in fact last longer than 1-1/2 generations and have walls somewhat better than the cheapest residential construction, some school architects nonetheless designed to the government-dictated lower standard, believing they were protected by the inclusion of a reduced lifespan in the stated design program. Other school architects simply stopped designing schools for several years.

When these newer schools designed to lower standards began to fail, guess what? Government sued all of the architects (as well as contractors and engineering consultants), expressing astonishment that they would have embraced such low standards. At last count more than 300 schools had been extensively repaired. Standards were reinstated and some former school architects who had refused to work to lower standards were back in that business. As for the contractors and consultants who were sued, they settled out of court for undisclosed amounts and with no admissions of guilt. In a bizarre twist, some even resumed work for the school boards that had sued them!

The moral of that portion of the tale is simply to trust your gut if you are asked to reduce your standards. Remember that the room will empty quickly of supporters if there is ever a problem.

The mechanics of value engineering/design assist processes are sometimes spelled out in contracts between the client and contractor, which information may not initially be communicated to the consultants. For example, some construction contracts rebate a percentage of any cost savings to the contractor. The first time the subject arises, it would be wise to ask what other consultants' contracts say about the subject. Also to table the fact that consultant effort associated with re-researching and re-designing the modifications is an additional service. If there is resistance to explaining the process or how the net savings are evaluated, record that resistance in your meeting notes and circulate them to your client.

What are the principles & best practices? Value engineering proposals are often tabled as meeting minute items, or via email. This may lead to confusion about status, assignment, etc., especially as these items frequently have a short time frame for consideration. I have witnessed this confusion countless times and my suggestions come from that experience.

Hypothetically, a project team member, usually a contractor, proposes a design change, usually but not always to save money or schedule time, or both. Proposals need to be evaluated in terms of effects on the design and what planning and building authorities allow. Then proposals are costed if they still have merit. A decision is made based upon the evaluation. Documents are amended accordingly and the project proceeds. It's seldom as simple as what I've just described.

Occasionally, a proposed "savings" may seem obvious resulting from a simple substitution of one tile for another, as an example. However, many value engineered changes are much more complicated. After collateral costs are included, such as abutting material adjustments, consultant redesign time (read additional fees), amended permits, etc., the net savings may largely evaporate.

Consultants need to be paid for their effort analyzing proposals and making any consequential changes to design or construction documents. This is especially important as the short time frames associated with many value engineering proposals increase the designers' likelihood of making errors – the considered way that a design evolves may be abandoned under the pressure to save time or money. To guard against this, senior/experienced staff, typically the most expensive, should evaluate value engineering proposals and the consultants should be remunerated accordingly.

Clarity is the team's best tool. Since each proposal is essentially a standalone item, it

D	B	Subject & Notes: see Note 1 below
	x	Initiated by:
	x	Disciplines/Trades Impacted:
	x	Proposal Description:
		Design commentary:
	x	Construction Manager Comments:
	x	Quality Impact:
	x	Schedule Impact:
	x	Cost Impact:
	x	LEED Impact:
		Action Items:

Figure 41 - Value Engineering / Design Assist i-A

is best handled by an i-Action (i-A) that includes this content:
In my experience, several of the ten considerations above are usually missed on the first go around, especially design commentary and the consultant fees part of cost impact. If the prime consultant opens a value engineering i-A each time one arises, fills in the information available and immediately distributes it as a work in progress to the project team, then either the outstanding information will be provided or the prime consultant will be able to identify the item as incomplete.

Regardless whether you are the prime or a subconsultant, be sure to include your estimated design team additional fees, and if it is premature to put a $ figure to them, insert a placeholder like "Consultant fees to be advised when scope of work is clarified."

Having defined the breadth of information needed to value a value engineering proposal, the prime consultant is better able to obtain that information, leading to better estimation of consultant level of effort, hence remuneration and real net "savings" of the proposal.

RECORD the journey	Record each proposal as soon as it is tabled	**Summary Principles & Tools**	
RESOLVE the issues	Ensure all ten elements of a proposal are evaluted	**i-WorkPlan (i-WP)**	
REVIEW the results	Determine the real net "savings"	**i-Actions (i-A)**	One of each value engineering proposal
REMEMBER & learn to improve		**i-KnowHow (i-KH)**	

Links: To review a copy of the Value Engineering Report, go to http://bit.ly/aagc06-1

## 07

Bidding
"Mandatory" Site Tour

"Sorry, you missed the mandatory site tour, I can't accept your bid." – Architect/Prime Consultant to Bidder at bid closing time

What's the point? Builders think differently than designers. Even when a design team has, for example, included areas of a site for material delivery and lay down, there is a multitude of considerations that will have been missed: Where can our workers park? Who will keep the school kids and their parents away from our concrete trucks?

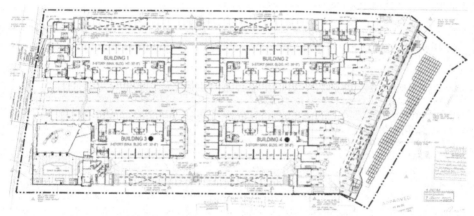

Figure 42 - Where is the lay down area?

A site tour held during the bidding period will resolve many of the bidders' questions. Many clients insist attendance at such a site tour is a precondition to being able to submit a bid. You should too.

What are the principles & best practices? The majority of tender processes include a site tour, usually about one week into the tendering period. By the end of the first week, serious bidders will have obtained tender documents and started looking at them. They will have contacted key subcontractors such as for mechanical services. They will have begun to generate a list of questions that they cannot find answered in the tender documents. The schedule for a site tour is often included in the tender documents; where it is not, it should be included in the first addendum issued by the architect.

The architect usually chairs site tours. Basically, everyone convenes at a location on the site, attendance is noted and introductions made and the group is toured around the site, asking questions as they go. It is advisable to include in the group a representative of the client, as there are usually some questions best answered from that perspective.

To get the best bids for a project's construction it is important to reduce the risks associated with bidding. Each time a bidder has a question that is not well answered, he/

she will add an allowance for the unknown condition. If documents are not well developed and coordinated, these risk costs will be significant, perhaps even endangering a project's financial viability. The principle of fairness also applies here. It is important that each question that is answered by the design team is answered for all, not just the person who asks.

In addition to fairness, the reason for closing the bid list to bidders who have not attended the mandatory site tour is primarily to reduce course of construction claims for "unforeseen circumstances". Where a contractor misses the site tour but is allowed to submit a bid, there is a greater likelihood of claims arising during construction: "I assumed I could lay down my materials there…"; *"I can't afford to hire a crossing guard, it's not in my bid…"*; etc. Also, calling a site tour "mandatory" in bid documents, then subsequently allowing other bidders who did not attend to submit bids, opens you to legitimate claims from compliant bidders that the bids from non-attendees are non compliant, being contrary to the bid documents, which had required attendance at the site tour, and should have been disqualified. Yikes!

Many public and private clients insist on site tours being mandatory, to reduce risk as described above. I have conducted many and agree, suggesting they are appropriate for all but the simplest projects.

The best way to set up a mandatory site tour is to be clear about it in your bid documents: *"There will be a mandatory site tour on [date] at [place]. Bids will only be accepted from general contractors who have attended this site tour. Contact [architect's office] to register. General contractors may bring subcontractors with them, but subcontractors may not attend instead of a general contractor."* Another example follows (the [square brackets] are part of my spec writing software and indicate options to select from): Consult with your client about best timing for a site tour. If the project is impacted by, for example, pedestrian travel routes in and around the site, it might be a good idea to schedule the meeting during one of those busy times, as additional public safety measures may become obvious and necessary. Even if they are generally

E.	Examination of the Site
1.	Prior to submitting tender, Bidders shall familiarize themselves with conditions and limitations that may affect the performance of the Work, including related bylaws, and regulations. Bidders may make arrangements to examine the areas of work by contacting the Owners.
2.	Bidders are **[required to]** **[invited to]** attend a **[mandatory]** pre-bid site meeting to review the building and the scope of the work. The Consultant will schedule the meeting.

Figure 43 - Mandatory site inspection in Bid Documents

covered in the specifications, there will be no excuses if bidders are exposed to what is going on around the site.

The meeting chair should come to the meeting with a "scribe," someone from the office whose job is just to take notes of what is asked and answered. This is a good mentoring moment for younger staff. The scribe can also help carry a copy of the bid documents (unless they are on a tablet computer as described elsewhere in this e-book). Ensure the main consultants also attend the meeting, i.e., civil, structural, mechanical, electrical, landscape; others if the nature of the project merits. These consultants can also carry a copy of their portion of the bid documents.

If your site plan includes areas set aside for contractor use, start in those areas so that attendees can immediately see what is available to them, identify access routes, etc. Where your project is an addition or renovation, escort the group through the areas to be renovated, also bring them to the location(s) where additions are to join with existing structures. There are usually lots of questions in these areas of complexity, ranging from the provision of services during construction, to maintenance of safe exit routes for building occupants, etc.

As you conduct the tour, you will be asked many questions, every one of which the scribe should capture. Where you, a consultant or the client representative can answer on the spot, do so, noting both the question and the answer for later inclusion as a clarification in the next addendum. Where you cannot immediately answer a question, say so and indicate the answer will be contained in the next addendum.

After the tour has been completed, keep the design team and client representative behind, to confirm who is to answer which of the questions that are outstanding. The bid period is short, so these questions demand immediate attention. Give a deadline; 24 hours should generally be sufficient. Give the team both your and your scribe's contact information, in case one of you is in a meeting or otherwise indisposed. Insist on written answers to the questions. Where a design team member needs to clarify the bid documents as a result of the question, ask that the answer, be it a sketch or specification change, be drafted in addendum wording (more about that later) so that you can efficiently prepare an addendum.

Assemble the site tour results as an addendum consisting of: meeting minutes covering those questions asked and answered; a post meeting memo answering those questions not answered on the spot, which do not require changes to the bid document; and an addendum covering answers to questions asked that have resulted in a change or clarification to the bid documents.

Issue the site tour addendum to all general contractor attendees, the design team and client representative.

RECORD the journey	Use a scribe to capture all comments and questions during a site tour	Summary Principles & Tools	
RESOLVE the issues	Cover off all questions asked during the site tour	i-WorkPlan (i-WP)	
REVIEW the results		i-Actions (i-A)	an i-A for each subsequent addendum item
REMEMBER & learn to improve	Modify specifications and bid documents to cover off items identified as not clear during site tours.	i-KnowHow (i-KH)	

Links: Go to http://bit.ly/aagc08-2 for further information, details and formats about addenda.

Bidding
Refining the Bid Documents by Addenda

"My good friend Bob the General Contractor has a great carpet he wants to include in his bid. I know the bids close in two days, no problem. I just want the best deal, and Bob's always treated me right." – Client to Architect

"I understand he did not acknowledge receiving the last four addenda, but he is the low bidder!" – Client applying pressure to Architect

What's the point? Next to construction defects and payment certification, insurers and lawyers tell me that problems arising during tendering are the most common source of litigation in construction. Why?

Consider: for a traditional design consulting practice, the highest component of over-head is generally "the drafting office." Regardless whether construction documents are generated by hand or computer, this part of the design and construction process is usually the most costly part for the designer. Hence the higher percentage of fees generally associated with this phase.

Designers should realize that contractors, as well as many clients, have little idea what goes on in the drafting office. Most do not understand the process of design as designers know it, i.e., iterative, painstaking, full of judgment calls, etc.

For the contractor, the equivalent of the drafting office is the estimating group. Con-tractors are always estimating. They estimate bids for new projects. They estimate budgets for repeat clients or construction management projects. They estimate costs for changes to current projects. For the most part, the only income earned as a result of their efforts is a modest preconstruction fee or the 5-10% markup builders usually receive for approved changes. There is no bonus to the contractor when their esti-mates land a project – they just get to build the project, generally for a price lower than their competitors on that day. Their success rate for new projects may be as low as one in ten.

Not surprisingly, most designers have little idea what goes on in the estimating depart-ment of the contractor's office. They seldom understand the large number of contacts an estimator needs to make to secure a price that the designer sees as one simple number. They frequently do not understand that a design change might involve getting pricing from eight or ten different trades, all of who are busy in their own estimating departments.

So contractors are very keen to be successful in their estimating in order to remain in business and be profitable.

Estimators are very good at estimating the costs of what they know. They are also good at identifying what they don't know and attaching contingencies to the unknowns. Addenda reduce the unknowns leading to more accurate estimates with fewer contin-gencies. In this scenario everyone wins. The client gets best value for money; the de-signers get what they designed; the builder gets a fair return for a known work scope.

Understandably, contractors are not happy when the bidding process is upset. The two quotes above illustrate just two of the ways this can happen. In the first example, one bidder is given an unfair advantage by being allowed to provide an alternative product that may be unavailable to other bidders. In the second scenario, there is pressure to accept the low bid even though it may not be based upon a true picture of the work scope.

In these and similar circumstances, if you were a contractor and had the second-to-lowest bid you would not be happy. You will earn nothing for the exhaustive efforts of your estimating group. You may elect to file a claim against those you feel violated the bidding process. You will probably sue the client and the architect, perhaps others. And you will not sue for something simple like the difference between your bid and the low bidder. You will sue for loss of profit, loss of opportunity (the income you would have received to pay for your office, staff, etc.) and possibly more. And you may very well win.

To avoid such an outcome, the architect as prime consultant needs to conduct a scrupulous bidding process, especially in the management of emergent refinements that are usually captured in addenda.

What are the principles & best practices? During the course of bidding, any member of the design team, or the client, may be asked questions by bidders. Where these questions involve clarifying what is already in the documents ("Look at detail 6 on sheet A302"), there is nothing required of the design team beyond recording what questions were asked and what answers given.

Where a question underlines an error or omission in the bid documents, or a bidder proposes and the designer approves a product or material different from what is in the bid documents, an addendum is required to add or modify bidding information; it is required by all bidders in enough time that they may all consider it in finalizing their bids. The specifications that form part of bid documents are written in active imperative mood – "Do this, furnish that, complete by doing this", etc. This is the language builders are used to, and what they use on site. A Superintendent does not say to a foreman "The steel studs should be plumb to the standards of CSA...", he/she says "Replumb the studs, they are out of true!" And that's the polite printable version!

Addenda need to be just as clear as specifications. A typical addendum entry might look like this:

1. REFERENCE: Drawing A-202, Second Floor Plan, Room 217
2. ADD: Closer Type CT-1 to Door 217-1; Hinge type H-4
3. DELETE: Bomer hinge type BH-2
4. SUBSTITUTE:
5. CLARIFICATION:
6. REASON: Building authority advises Bomer hinges not acceptable in this location

Figure 44 - Typical Addendum Item

(In this instance, headings 4 and 5 have been left even though they are blank, to illustrate the 6 headings that might arise in an addendum item.)

For consistency within your practice, it is always best to have a standard order for addendum contents. The one I have found works best is:

1. Revisions to previous addenda first – these are the easiest to confuse for the builder, as in a previous addendum you said "Do X" and now you are saying "Do Y" instead, or "Do modified version of X".

2. Revisions to specifications including additional specification sections – In most contracts, specifications take precedence over drawings, so best to write your addenda in that order. Occasionally when you do this, you will realize that there is still a conflict between revised specifications and revised drawings. Better to catch it now!

3. Revisions to drawings including additional sketches – I am unsure why it is that what we designers love to do most is of least contractual import, but there you have it. I think it must have something to do with the fact that lawyers can't draw.

There should only be one set of addenda issued on a project, and only by the prime consultant. They should include content solicited from the Client and all Consultants. Presumably to save time, some architects permit, even encourage engineering sub-consultants to issue their own addenda. Don't do this! It may appear to save time, but will only lead to chaos and confusion as the bidders try to determine if they have architectural addenda A-1 through A-4, Mechanical addenda M-1 through M-3, etc.

Do not wait for the consultants to volunteer. Communicate! Say *"An Addendum is going out Wednesday morning so I need your content by Tuesday noon latest."*

When you issue an addendum, issue it to every bidder, all of the Consultants and the Client representative(s).

At the start of the bidding period, establish with the design team a cutoff date for addenda content. You need to allow enough time before bid close for the bidders to: read the addendum; circulate it to their trades and suppliers; get quotes back from those folks; select which pricing to proceed with; and update their bid. This may be a 3-day process for simple items, more like five for others.

F.	Clarification
	1. Bidders finding discrepancies or omissions in the tender document, of having doubt as to the intent, shall submit a written request for clarification to the consultant at least [4] working days prior to tender closing.

Figure 45 - Addendum Cutoff Date Established in Bid Documents

If it emerges that there is a major design change required, for example to the HVAC systems, consider delaying the close of bids. The alternative will most probably be inflated bids because the bidders will quite rightly add a significant contingency for their inability to adequately price the change in time for bid close.

I make it a habit to be proactive, contacting the bidders early in the last week of bidding and taking the pulse. If most bidders, for example, say they are unlikely to be able to get good mechanical pricing by bid close, and can reasonably explain why, that may be reason enough to extend the bids. Conversely, if only one bidder expresses concern about the close date, that is probably insufficient reason to extend the bid period.

RECORD the journey	Track all queries during bidding	**Summary**	**Principles & Tools**
RESOLVE the issues	Ensure all bidders are aware of all modifications	**i-WorkPlan (i-WP)**	
REVIEW the results		**i-Actions (i-A)**	Use i-A's to capture specific addenda items
REMEMBER & learn to improve	Modify office standards to capture new knowledge from addenda	**i-KnowHow (i-KH)**	

09

Bidding
Accepting & Opening Bids

"You can't use any of the fax machines until after 3 p.m." –Estimator to Contractor staff.

During the recessionary times of 2008-09, I was with a contractor who successfully bid on a school building. On the day of bid closing, the estimator in charge stuck his head in every office on the floor and said politely, "You can't use any of the fax machines until after 3 p.m." Sensing a high level of trade interest in the tough economic times, he had assigned a junior estimator to every fax machine on the floor and given many small groups of trades their own fax machine number for bid submission, affording many more communication venues to the bidders. After we were successful on the project, a majority of our successful subcontractors (all had been prequalified) congratulated us, noting, "You were the only general contractor who received our quote – everyone else was jammed up!" in addition to the ongoing overhead of our estimating department, we had arguably incurred even more costs by assigning staff to man the fax machines – BUT WE GOT THE JOB.

What's the Point? The bidding process is a major component of every builder's overhead, so must be conducted as openly as possible to demonstrate to bidders that their significant and costly efforts are being fairly reviewed. Ideally, the architect/prime consultant should open bids publicly. I would go so far as to say that where a client opts to open bids privately, the architect should advise against that process in writing, or at least confirm in writing that she/he is acting on the owner's express direction.

B. Deliver Tender and Tender Documents to the **Consultant's** office by <time> p.m. on <date>.

C. Tenders shall be valid for a period of[30] [60] <specify length>days.

D. Award
   1. The Owner reserves to right to reject any or all tenders.
   2. The Owner reserves the right to negotiate changes with lowest bidder and his named subcontractors.
   3. The Contractor notified of acceptance shall, within [5] working days, execute and deliver CCDC-2 to the consultants office.
   4. The Contractor notified of acceptance shall, within [5] [10] <specify # days> working days, actively commence the work.

Figure 46 - Typical Tender Delivery & Award Considerations in Bid Documents

What are the Principles & best Practices? As bids are received, the receiver writes the time and date received on the outside of the envelope, but does not open it. If anyone walks in after the closing time, his or her envelope is not even accepted (That may sound corny but will become obvious as you read on).

Typically the bidders' representatives will have been at your door a half-hour or more in advance of the bid closing time, awaiting a call, email or text on their smartphone. Their instruction will be simply to fill in a tender amount that has been assembled up to the last minute by the estimators back in the office. This is all legit.

61

After the bid closing time has passed, the formal thing to do is to invite the bidders' representatives into the meeting room where you are opening the bids. As you open each bid, the script is something like this:

"I have a bid from Acme Construction Ltd. in the amount of $12,754,000.00. Their bid form confirms receipt of Addenda #1 through #4. They have enclosed a 10% Bid Bond and Undertaking of Surety for a 50% Performance Bond, as requested. The bid form has been signed and does not identify any exceptions."

Meanwhile the bidders are all madly writing down these details, to see where they stand as against their competitors. When you have opened the final envelope, your closing script might be:

"The lowest bid for the project appears to be $11,995,000.00 submitted by Beta Construction. All of the bids will now be reviewed in detail by the design team, the Client and the Client's lawyers and insurers. We will make our recommendation to the Client. You should expect to hear from us within the specified acceptance period. The selected bidder will receive a Letter of Intent followed by a contract. Thank you for your efforts."

Notice I have not committed to Beta. There is usually a clause in the bid documents saying "The lowest or any other bid may not be accepted." This clause is primarily designed to allow the Client, on lawyer's advice, to disqualify a bid based on circumstances that I as a non-lawyer cannot discern. I once had a project where every bid was non-compliant for a variety of reasons. Fortunately the client's lawyers determined that the lowest bidder's non-compliance was no worse than anyone else's, and on that basis he recommended accepting that bid. There were no ramifications, but there might have been.

Disqualification of a bid does not happen often. Where it does occur, it is the client's and their lawyer's responsibility to identify why, and to deal with any push back from any disgruntled bidders. If you have conducted your part of the bidding process properly, you should be fine.

RECORD the journey	Note the time of receipt on each bid	Summary Principles & Tools
RESOLVE the issues	Note that the owner's lawyer and insurer will review the bids	i-WorkPlan (i-WP)
REVIEW the results	Review the details of each bid	i-Actions (i-A)
REMEMBER & learn to improve		i-KnowHow (i-KH)

## 10

Bidding
The "Not Awarded" Letter

" I thought Beta would get the job based on their bid, but it's nice that the architect confirmed it to me – he's always so up front about the process." – Unsuccessful bidder to colleague

What's the point? Just as the design firm that comes second in an interview gets no tangible benefit from their efforts, so too the contractor whose bid is "second low."

What is the best practice? Consultants cannot and should not interfere with the bidding process. But that does not prevent them from thanking the unsuccessful bidders for their efforts – it will be appreciated.

2014-04-29
Project # [#]

[Company]
[Address]
[City, Prov PC]

Attn: [Name, Title]

Dear [**name**] :

Re: [Project(s)] – Award of Contract to Others

Thank you for your bid for the above project. We would like to advise you that the contract has been awarded to [Insert name of successful General Contractor] [for an amount of $[Insert contract amount]]

[Would you please return the bid documents to our office at your earliest convenience.] [Insert information regarding return of bid security or bid deposit after execution of contract, if necessary.]

Thank you for bidding on this project.

Yours very truly,

Figure 47 - Template for "Not Awarded" Letter

RECORD the journey		Summary	Principles & Tools
RESOLVE the issues		i-WorkPlan (i-WP)	
REVIEW the results		i-Actions (i-A)	Template(s) for unsuccessful bidder letter
REMEMBER & learn to improve	Thank the unsuccessful bidders	i-KnowHow (i-KH)	

LINKS: The "not awarded" letter can be found in the sample project i-WP under http://bit.ly/aagc10-1

## 11

Meeting in the Field
Begin as you Mean to Continue

"If the Owner, Architect or Contractor are not made aware of a design or construction change, then it never happened, does not exist and will be remedied at no cost to any of those three." – Architect/Prime Consultant to all First Site Meeting Attendees

"I'm glad you asked that question." "You always ask the questions I'm too shy to ask." "I had no idea who half the folks in the meeting were – thanks for asking for introductions." – various first site meeting attendees

What's the point? Site meetings are the one predictable, repeat occasion when all of the key construction phase players meet – the owner's representative, the contractor, the consultants and key subcontractors. Progress is discussed, issues are raised, debated and (hopefully) resolved. Site meetings can be used to establish expectations, then to monitor their achievement.

What are the principles & best practices? The construction stage of a project is run by the builder, including site meetings. Of course, designers are essential team members, but all construction contracts clearly place the builder in control of construction. However, that does not prevent the other team members from ensuring their reasonable requirements are included in the mix.

Project construction should include an initial site meeting fairly early in the process, typically shortly after the builder is established on site with shelter, power and communications. If the initial startup meeting is unreasonably delayed, that may be a symptom of bigger problems, so expect and insist in writing on an initial site meeting early on.

Lack of facilities is seldom an excuse. Most specifications are quite detailed about what temporary facilities are expected to be provided, typically: a site trailer for meetings (and builder offices), furnished for meetings, heated in winter, cooled in summer; temporary power to operate office basics such as a photocopier and fax machine (yes, contractors still use fax machines); temporary washroom facilities; references, including contract documents, applicable building codes and sometimes other standards. A site telephone used to be standard but has largely been supplanted by everyone's cell phone. Wi-Fi connectivity is becoming more commonly specified.

Site meetings are almost always chaired by the builder, which is great for the designers not because they have less work, rather because it reveals the mindset of the builder – what is emphasized, what is glossed over, what is considered important, what is incidental?

I have made it a habit for years to bring my own standard agenda to first site meetings. As items are covered off, I make appropriate notes. At the completion of most site meetings, including the initial meeting, everyone around the table is asked for their "new business". I then bring up anything on my agenda not covered by the chair or other contributors. I have never had a builder object to this; they are usually happy that something they inadvertently omitted was raised by others and covered off. There are no dumb questions or observations.

I use an i-Action (i-A) for the initial site meeting agenda. It typically includes the following – note that some of this may seem obvious, but it is seldom all evident at the beginning:

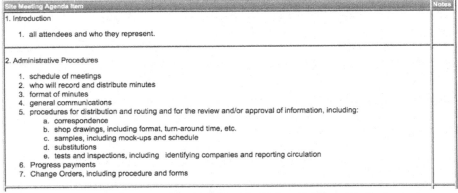

Site Meeting Agenda Item	Notes
1. Introduction 1. all attendees and who they represent.	

2. Administrative Procedures
 	
   1. schedule of meetings
   2. who will record and distribute minutes
   3. format of minutes
   4. general communications
   5. procedures for distribution and routing and for the review and/or approval of information, including:
      a. correspondence
      b. shop drawings, including format, turn-around time, etc.
      c. samples, including mock-ups and schedule
      d. substitutions
      e. tests and inspections, including identifying companies and reporting circulation
   6. Progress payments
   7. Change Orders, including procedure and forms	

Figure 48 - First Site Meeting Agenda - Getting Started

Have you ever been in a meeting where you did not know who half the attendees were, better yet what they did? If it's not evident, be the guy like me who raises his hand and says, *"Sorry for asking the dumb question, but can everyone introduce themselves and indicate their contribution to the project. I'll start..."* After the dust settles more than one person will thank you for asking the question.

There are many meetings involved in construction, typically: periodic site meetings involving the Owner's representative, the Architect and Consultants, and the Contractor (OAC meeting), maybe monthly at the start of the project when there are few trades, perhaps biweekly or weekly as the pace picks up; subcontractor meetings involving the General Contractor or Construction Manager together with most or all of the subcontractors on the project, also often including major suppliers such as the concrete supplier; superintendent meetings on larger projects, where the Superintendent team gets together, often daily, to schedule work, table issues, etc.; and contractor project team meetings, where all of the builder's site personnel and sometimes head office personnel meet to discuss the general state of the project and issues that involve all interest groups.

At one of the best-run major projects I have ever participated in (as a contractor employee), the entire contractor project team met at the start of every workday to briefly review status, challenges for to-day, etc. The meeting seldom lasted more than twenty minutes. What was perhaps most impressive for me was that the vice president responsible for the branch was almost always present and an active participant.

All of these meetings talk about schedule and issues together with other agenda items particular to the audience. Consultants and the Owner are typically only welcome at the OAC meeting, but it is helpful for them to have the sense of the other meetings scheduled so they can mesh some of their efforts with the contractor's more frequent assemblages. If you know the superintendents and subcontractors meet every Tuesday morning, you can call the Superintendent on Monday to say, "Can you ask the window sub when I can expect to see those shop drawings, please." That's not being intrusive, just efficient.

The builder typically records minutes of all meetings on construction sites. For the Owner and designers, this is a valuable look into the builder's universe of thoughts, processes, etc. Read meeting minutes carefully and quickly identify anything that you remember differently – the difference is likely innocent and honest, but needs to be clarified. Express any concerns in writing to the minute taker in the time frame noted in the minutes, usually 24-72 hours. This does not mean you might not see something in the minutes later on, in which case you raise it immediately. But if you regularly dispute meeting notes too long after the fact you will only damage your own credibility and reputation.

Regarding general communications, in this day and age email (and in my case, i-A's) will suffice except for a few formal contract notifications usually involving major problems. Contracts vary between countries and jurisdictions within countries. And contracts customized by Owners, designers or builders may have different rules. Read the contracts carefully and make notes in your i-WorkPlan about unusual communication or notification requirements.

At the outset of a project it is important to have a complete team list. If you are using my recommended three tools, this team list will be a printout from your i-WorkPlan database rather than a standalone specially created document

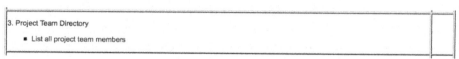

3. Project Team Directory
  ▪ List all project team members

Figure 49 - First Site Meeting Project Team Directory

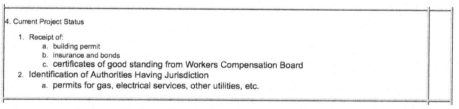

Figure 50 - A Project Directory from a fictitious Project i-WP

Ordinarily the Project Directory will be many more names than this. This printout from QW is for illustrative purposes only. My largest project to date had i-WP's with 800 members.

4. Current Project Status
  1. Receipt of:
      a. building permit
      b. insurance and bonds
      c. certificates of good standing from Workers Compensation Board
  2. Identification of Authorities Having Jurisdiction
      a. permits for gas, electrical services, other utilities, etc.

Figure 51 - First Site Meeting - Current Project Status

It is important to identify the project's current status, not always evident from the state of the construction site. I am unaware of any project in any village, town, city, county, state or province that can start work without a building permit. Actually, that's not quite true. I once owned a cabin on Hornby Island in the Strait of Georgia, where adherence to the building code was described in official documents as "optional".

Verify with your own eyes that the necessary permits exist, which is easy in many jurisdictions, which require they be posted in a prominent location. It is almost always the builder's responsibility to obtain the building permit, but everyone gets in trouble if she/

he has not done so. If you are a self-governing professional (as many architects and engineers are), your duty to protect the public requires that if there is no building permit you should advise the contractor that work should not proceed until there is one. If this is challenged in any way, you must leave the site, head straight to "city hall", report the circumstances, then duck as the fur begins to fly. You cannot and should not stop the work – that is a local jurisdiction prerogative, and they will do it. But if work is allowed to proceed without permit, you are not protecting the public interest. I have been the whistle blower on a handful of occasions in my career. It lost me one client, but I later learned being fired by him was the preferable alternative, the other being an inevitable lawsuit in lieu of fee payment.

On certain projects such as "fast track", and in certain jurisdictions, local authorities MAY issue partial or sequential permits, e.g.: excavation and shoring; then foundations up to and including grade; then construction above grade; etc. But the principle of no start without a permit still applies. This can be particularly challenging when the second or third sequential permit remains elusive while the pace of construction threatens to overrun the scope of the issued permits. I have on several occasions advised a contractor to stop work until permits for the next phase were in place.

Bonds and insurance are almost as serious. These are between the builder and the Owner. If they appear in evidence, always recommend in writing that the Owner have them reviewed by their lawyer and/or insurer – you are neither of these and cannot determine if they are complete or accurate. All the designer is expected to do is determine if the correct types of bond appear to be in evidence, and that the amounts appear to be what has been specified.

If either bonds or insurance are absent or incorrect, a project that has started is in immediate jeopardy. Insurance covers costs (minus deductibles) of the specified things that can go wrong, from loss of use through loss of life, damage to property, injury to people, etc. Most workers, suppliers, material delivery people and of course the general public will assume if work has started then insurance must be there. If you are aware of gaps that should be filled, and carry on with your work, you may be considered negligent in the event of loss or damage. Your own liability insurance may protect you somewhat (minus deductibles, which are usually hefty), but it will not cover loss of reputation and loss of confidence by the Owner and the general public. It takes years to build a positive reputation, moments to destroy it.

Some years ago I was working on an addition to a school. At the first site meeting, I asked about insurance and was assured all of the builder's insurance was in place. When I asked if the school board had advised its insurer that an addition was underway, there was silence around the table. At my suggestion, the owner's representative conferred with their insurer immediately after the meeting. As it happened the school board's insurance policy did not cover its potential liabilities during construction. An inexpensive rider was quickly added to the insurance policy and work proceeded without incident. I was thanked for my diligence. I am no insurance expert, but in principle, insurance is intended to cover events that occur during the normal course of an insured's business. Construction is normal course of business for a contractor, for an owner maybe not so much.

Another way to regard all this paperwork is as what I call "Project Documentation Submittals." My master specifications include a list of all of the paperwork I have ever had to provide in one table. I edit it down based on the project particulars.

ITEM	When due	Type of Submittal ("x" = req'd) BOND	INSURANCE	LETTER / Report	Sched-ule	Warranty
Performance Bond	10 days	x				
Labour & Materials Payment Bond	10 days	x				
Schedule of values	10 days				x	
Predicted cash flow requirements	10 days				x	
Letters of good stand-ing	?					
Sample stat dec.	?				x	
List of signing authori-ties	?			x		
Submittal schedule	?				x	
Schedule of rough-in dates for Owner-supplied eqpt.	?					
Mechanical eqpt. list	Prior to order				x	
Electrical eqpt. list	?				x	
Test reports	as/when			x		
Copies of permits, li-censes & certificates	as/when					
Regulatory Authority approvals	At sub compl					
Independent Authority approvals	At sub compl					
As-Built drawings	At sub compl					
Record Drawings	At sub compl					
WCB letter of good standing	At sub compl					
Reconciliation of changes to the Con-tract	At sub compl					
Local authority occu-pancy permit	At sub compl					
Deficiency list	At sub compl					
WCB letter of good standing	At total perf.					

Figure 52 - Almost Every Paperwork Submittal You Could Ever Imagine!

This list also gets distributed at the first site meeting and I resurrect it throughout the course of construction when appropriate.

When questions are asked, issues arise, etc., especially from local authorities, it is

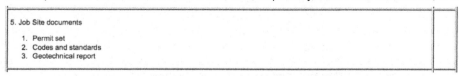

5. Job Site documents

1. Permit set
2. Codes and standards
3. Geotechnical report

Figure 53 - First Site Meeting Agenda - Job Site Documents

important to have on site access to the documents that permits were issued for. Doc-uments often disappear as they are grabbed "for a minute", so on most sites the lead Superintendent will have permit documents in a safe and secure location. These are the documents the local building and trade inspectors will be looking for to support their own work.

As with permit documents, so too we need access to the basic codes and standards that underlie the proposed construction. These might be paper or electronic documents, but they need to be available for periodic reference. Having access to all standards referenced in construction documents might seem a bit far-fetched, but as a minimum on site access should be available for standards that are key to the project. As an example, I recently worked on a project requiring a "super flat" floor slab. When tolerance issues arose, the standard was unavailable and by the time it was obtained there was an argument instead of a discussion.

If applicable codes and standards are addressed at the first site meeting it is more likely that those that are key to project specifics will be available if needed.

As with the other first site meeting agenda, note action items, as well as inaction items. If, for example, you have recommended that the owner's insurer review the project's insurance coverage and your advice is declined, you want a record to that effect.

The items under site issues above may seem inconsequential, even annoying, but all

---

7. Site issues

   1. Parking
   2. Loading and storage
   3. Garbage, construction waste and recycling
   4. Hoarding and fencing
   5. Tree protection
   6. Project identification and signage
   7. Snow removal
   8. operational constraints (use of existing   washrooms, maintenance of existing   services, etc.)
   9. hazardous materials (handling and   disposal)

Figure 54 - First Site Meeting Agenda - Site Issues

have the potential for problems. If construction forces disrupt neighbourhood parking, if streets are blocked for deliveries, if the site is poorly fenced, hence unsafe, problems will ensue. Green building projects will suffer if garbage and construction waste are not well handled, not to mention handling and disposal of hazardous materials. Many project's neighbouring buildings will be close to your excavation, or will be old

---

9. Requirements for Pre-construction Surveys

   1. Recording existing conditions
   2. Setting out foundations

Figure 55 - First Site Meeting Pre-construction Requirements

and perhaps not well built. Contractors (not consultants) should always photograph all neighbouring structures, hard and soft landscaping before any work starts, in order to have a comprehensive record of existing conditions. Crack lengths and widths should be measured and the measurements themselves photographed (i.e., photograph the tape measure). This is the Contractor's responsibility but worth discussing at the first site meeting.

Your own initial site meeting agenda will evolve over time, as has mine. What is essential is to have an agenda for reference and to ensure its contents are covered off.

Every project should have a schedule. The simplest project will run off a hand drawn

6. Review of information from Contractor, including:

    1. Construction schedule
    2. Schedule of values for progress draws
    3. Cash flow projections

Figure 56 - First Site Meeting Agenda - Time and Money

bar chart listing basic subjects like "excavation/ footings/ foundations/ slab/ framing/ roofing/ cladding/ doors and windows/ interior finishes." The most complex projects will have thousands of interlinked events requiring full time updating by highly trained staff.

One concept that is sometimes difficult for builders to understand is that their responsibility for the construction schedule does not impose a commensurate duty onto the designers. The smart designer will study the contractor's schedule and book time for reasonable field review in association with scheduled close-in's. mockups, installations, etc. But there is no real or implied contract between the builder and the designer that requires the designer to abide by the builder's schedule.

In most construction contracts, the designer's requirements for field review in relation to the builder's schedule is limited to words like *"intervals appropriate to the stage of construction that the Consultant, in his or her professional discretion, considers necessary to become familiar with the progress and quality of the Work and to determine that the Work is in general conformity with the Construction Documents."*[1] Say what?

Some designers abuse this horribly and almost never show up on site until near completion, when they designate many items as "deficient" that would not have been had they looked at them in a timely fashion. This is very unprofessional behavior, to say the least.

Conversely, builders will sometimes try to use designers instead of establishing their own quality control, calling for multiple reviews of each room on each floor in each building and expecting a room-by-room deficiency list throughout the course of construction. This, too, is very unprofessional. When this appears to be the case, the Consultant should discuss the abuse of services first with the Contractor, then with the Client if necessary. Assuming you are able and willing to provide more detailed review, it should only be as an additional service for an additional fee.

I have made it a practice for years to only provide field review on an hourly fee basis. I am happy to declare a reasonable budget for field review, and monitor my fees against it. In cases where my services are being abused, this is usually evident early on, allowing me to alert the client in a memo attached to my invoice.

In some jurisdictions it appears (meaning I cannot believe it!) that Consultants are able to legally contract out of any field review, although this seems open to local interpretation. In my experience, it is important to provide field review regardless of the local minimum standard, else how can you hope to have your design accurately executed? More about this later.

---

[1]   RAIC Document Seven, Article 1.7 General Review / Field Review – other consultant contracts are similar.

RECORD the journey	Create and maintain a standard 1st site meeting agenda.	Summary Principles & Tools	
RESOLVE the issues	Identify & work to resolve missing elements	i-WorkPlan (i-WP)	Provide for initial & subequent site meetings
REVIEW the results	Track unresolved issues to completion	i-Actions (i-A)	for every emergent issue
REMEMBER & learn to improve	Continually refine your standard site meeting agenda	i-KnowHow (i-KH)	

LINKS:The first site meeting agenca can be found in the sample project at http://bit.ly/aagc11-1

12

Meeting
Resolving Site Meeting Issues

"The same issues resurface at every site meeting and never seem to get resolved. It's nuts!" – Just about anyone

What's the point? Almost every project has a few thorny issues that do not lend themselves to easy solutions. So they fester like an open wound that refuses to heal without a considerable application of effort. If one such issue arises at each bimonthly site meeting, before long there will be a dozen of them and they will consume increasing site meeting time as well as sidebar emails, phone calls, etc.

It is easy to ignore the cost of these unresolved issues because most meeting attendees are on salary or being paid a fixed fee. The true cost of a site meeting can easily surpass $2,000.00 per hour when you calculate the time cost of clients, contractors, subcontractors and consultants. So those unresolved issues become very expensive very quickly.

What are the principles & best practices? The solution is simple but elusive. Each emergent issue should be an i-A assigned to one responsible party, with one other person charged with helping the assignee get to solution. A due date should be assigned, which in most cases should be before the next scheduled meeting so that the results can be reported and the item closed.

Identifier	Number	Floor	Item	Due Date	Opened	Status	Priority	Type - Subtype	Company Name - Now Responsible
Project Risk	A00127 R0		Proposed new Wall cladding material	2014/10/01	2014/09/22	Open	04 by next meeting or Due Date	Issue -	Copyright 2014 Brian Palmquist - Brian Palmqui

Figure 57 -Every Issue Assigned to one Person

If you grab onto the items that are truly yours to resolve and set a target before the next regular meeting, two things will happen: 1). You will become more efficient and profitable; and 2). The other players, including the client and contractor, will come to know that you are a problem solver. There is no down side to having such a reputation.

I have been wearing my quality management hat in all aspects of practice for some time now, because a quality-managed approach always produces better results with less effort and greater profit. In the issue resolution arena, all assignments become i-Actions. Their content has many forms such as checklists, boiler plate, tables, etc., but they always have these elements that are fundamental to quality-managed practice: a unique identity and tracking number; a priority; a due date; "referred by" and "now responsible" team members. There are other optional filters as well.

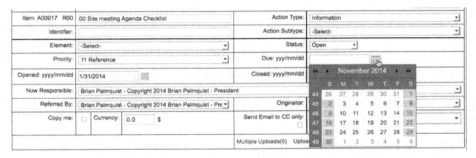

Figure 58 - i-A -Assignor, Assignee, Due Date, etc.

Consistent with quality-managed practice, when an i-A is assigned via email, a record of its issuance with all details is automatically created in case needed later. All attachments, replies, etc., that arise are automatically aggregated to the originating i-A. When it is deemed by the assignee to be closed, the closure can be confirmed via email to team members who need/want to know, and a record of that closure event is also automatically created and attached to the i-A.

RECORD the journey	Create & store most records automatically	Summary Principles & Tools	
RESOLVE the issues	Each issue has a separate i-A history	i-WorkPlan (i-WP)	
REVIEW the results	How do we measure how we have done?	i-Actions (i-A)	Specific forms or templates
REMEMBER & learn to improve	Convert i-A's to i-KH's where new knowledge emerges	i-KnowHow (i-KH)	

LINKS: For a more detailed explanation about i-A's go to the sample project Actions list at http://bit.ly/aagc12-1 and click on the "Help" button at the top centre of the screen.

## 13

Meeting In The Office
When Are Meeting Minutes Not?

"Did I really go to university to learn how to take minutes of meetings?!" – Intern architect to principal architect.

"The same things get discussed at three different meetings in the same week and minuted at each meeting. The duplication is driving me nuts!" – Administrator to junior architect

What's the point? As a younger architect I was impatient. Impatient to get back to designing. Impatient to get back to the long list of "to-do's" on my desk. Impatient to do just about anything other than write notes about talking that would then be used to create more talking and notes, mostly duplicating the earlier talking and notes. Surely, I thought, there has to be a better way.

When are meeting minutes not meeting minutes? When they are i-Actions (i-A's). Think about what meeting minutes are. Once you get past the roll call, they are a recitation of certain categories of existing and new issues requiring action, i.e., i-A's. "Old business" is generally nothing more than the categories of issues such as non-conformances, submittals, RFI's, etc. that have not been closed yet. "New business" is emergent items of the same sort. All can be managed as i-A's.

For each type of meeting (site, RFI, Change Order, etc.), why not simply electronically assemble i-A's from a single database, focusing on how they have been "tagged" rather than where they have been filed or written – any combination of subject, type (submittal), subtype (reviewed as noted), assignee (contractor), due date, etc., should do. If the same i-A pertains to multiple subjects (for example, a RFI assigned to the mechanical contractor for Building B), it can easily be tagged so as to be a single data entry considered at three separate meetings until it is resolved, but never duplicated along the way, nor added to one, two or even three other meeting minutes.

With this approach, meeting minutes disappear as separate documents requiring copying and pasting, dual entry of data, etc. For the sticklers about history, a PDF version of agenda items, including further detail and resolutions, can be captured as a "document" and emailed to participants. But since each i-A will include its own history (Who created it? Who responded? Who attached information? Who closed it?), the need for standalone minutes will gradually disappear, especially if the meeting chair simply refers to i-A's on an LCD projector or smart board.

In between meetings, as items such as RFI's, deficiencies, etc., emerge, they can be set up as independent i-A's, assigned, worked on and possibly even closed. Fill in any missing tags such as assignee, due date, etc., type, subtype, etc., and they will simply appear as part of the agendas of appropriate meetings.

Many meeting minutes carry over items closed between meetings, simply because there is no other way using traditional meeting minutes to tell the audience that an issue has been closed. If your i-A structure prompts team members to advise when an i-A is closed, then this caterpillar-like dragging along of closed issues goes away, and team members are advised at the time of closure rather than at the next meeting.

RECORD the journey	Capture every issue, but only once	**Summary Principles & Tools**
RESOLVE the issues	Each issue is standalone and tracked once only	**i-WorkPlan (i-WP)**
REVIEW the results	Keep tracking each open i-A until it is resolved	**i-Actions** an i-A for each **(i-A)** issue
REMEMBER & learn to improve		**i-KnowHow (i-KH)**

LINKS: For a sample i-A "header" to lead off a meeting as it transitions to i-A's go to http://bit/ly/aagc13-1

## 14

Liaising With Local Authorities

"The building inspector on the last job had no problem with our code interpretation about the guardrails" – Architect to Client

"I can feel as I walk up this stair that the range of riser height is more than +/- 1/8", *which is unacceptable."* – Building Inspector to code consultant.

What's the point? Each municipality, sometimes each inspector in each municipality, has their own history with building code interpretation, often built on their own tales from the trenches. Although I have never been involved in a guardrail dispute, I am aware of accidents around guardrails after which lawyers measured their height, picket spacing, etc., and fought legal cases based on millimeters (1/16"). Similarly, "trip and fall" cases have been decided on what seem to be minor variations in riser height, tile joint lippage , etc. Unless an interpretation is just plain wrong, you need to work with the local authorities and their "hot buttons" to be successful.

What are the principles & best practices? Most building inspectors spend most of their workday inspecting buildings under construction. That may seem self evident, but due to insurance and legal claims there are now jurisdictions where building inspectors have been told only to ensure that professionals are responsible for field review, rather than provide any themselves. A building inspector who does not inspect buildings may sound bizarre, but is sometimes the reality. And where such a situation exists, it may place additional liability onto the professionals, since there is no regulatory inspection occurring, no "second set of eyes."

There is often no rhyme or reason as to what each building inspector inspects, so beware when a contractor says "...on our last job, the building inspector was fine with that." It's worth checking the rules for each project, and with the current inspector.

As a best practice, somebody with knowledge and without a conflict of interest needs to review enough work under construction to be able to judge that it is likely that all aspects of the building have been constructed "to code," in particular life safety systems such as fire and smoke alarms and systems, exits, fire separation, fire stops, etc. The contractor remains responsible for building from approved documents, but will largely delegate that responsibility to the subcontractors constructing their specialized portions of the project. The contractor's superintendent is the second set of eyes, yours are the third and the municipal inspector's are the fourth. Sounds like a lot, but in a high-pressure project (i.e., all of them), no subcontractor will likely see everything in their scope and no superintendent will likely verify every installation. Thus it will fall to the inspectors' and your periodic review to identify if work appears to be in general conformance. Chances are if you identify a number of issues during a periodic review, then there are others.

We talk in another tale about the value of preparing a list of minimum required field reviews, a document I give to the contractor as a reminder of when I expect to be called out. If you use the i-Action (i-A) approach, your list of Required Field Reviews might look something like this (in the figure below, A = Architect, C = Consultant, BE = Building Envelope, M/P = Mechanical/Plumbing, E = Electrical, R = Roofing, RA = Regulatory Agency):

A	C	RA	N/A	Minimum Required Field Review	Comments/ Conclusion/ Action - Minimum field review intervals
x		x		Foundations - Includes dampproofing, waterproofing, weather barrier under slab, below grade/slab insulation	Once per building or once per elevation
x	BE	x		Moisture content - wood frame - Requires bldg be enclosed with roofing, weather barrier & all doors/windows in place except at drywall loading locations.(i.e., should be entirely weather protected)	Once/bldg -> townhouses; once/floor others. Note, Contractor identify RA review requirements for framing.
x	BE			Window/Air Tests - Typically 1% of windows are tested, min. 2 per phase, one early and 1 late in phase. The later test may include air testing of a typical completed unit. Coordinate with Consultant which bldg(s) have windows tested.	Divide between phases or buildings; approx. 1/2 earlier, 1/2 later
x	BE M/P E	x		Pre-drywall - Requires insulation in place, and as req'd, ADA sealant and gaskets, window sealant, poly v.b., f'stopping, ductwork sealed, other special conditions, etc.	Once/bldg -> townhouses; once/floor others. Contractor identify RA review requirements for pre-drywall
x	R			Roof - Requires roofing & flashing complete - important to conduct early in project to establish performance requirements.	R = independent roofing inspector if spec'd;
x	BE			Exterior Pre-cladding - Includes preparations for cladding on all exterior wall surfaces, i.e., moisture barrier and strapping measures in place. Often combined with other milestone reviews depending on supt. organization of work.	Once/bldg -> townhouses; once/floor others.
x	BE			Exterior Envelope - Includes substantial completion of all exterior surfaces - may be combined with Pre-occupancy if circumstances permit.	Once/bldg -> townhouses; once/floor others
x	M/P E	x		Firestop systems - Submittal required describing each f'stop system proposed, for CP review.	Once/bldg -> townhouses; once/floor others. Contractor identify RA review requirements.
x	M/P E			Chases - fire rating pre-board - For lower fire ratings (up to 1h), typically requires 3 of 4 sides of chases be drywalled and taped/sanded and 4th side be ready for sealing. Verify method of sealing 4th side as acceptable per code/local building authority.	Once/bldg -> townhouses; once/floor others. Contractor identify RA review requirements.
x			x	Fireplaces - fire rating behind, pre-board - Requires wall areas behind fireplaces have insulation, vapour retarder and taped drywall in place.	Once/bldg -> townhouses; once/floor other. Contractor identify RA review requirements.
x	M/P E	x		Corridor ceiling drops - Ceiling areas insulated as required, taped with penetrations firestopped.	Once/bldg -> townhouses; once/floor other. Contractor identify RA review requirements.

Figure 59 - Portion of List of Minimum Req'd Field Reviews

With a list such as this the superintendent will know at the outset what your expectations are. In my experience, this leads to a matrix that the superintendent will refer to when looking for "sign off" so that work may proceed and areas be closed in without worrying about Consultants not seeing what they wish to. Nothing wrong with that.

As another best practice, there is no harm in visiting the building department of a jurisdiction you've not worked in before or for some time, introducing yourself and asking what the local building and trade inspectors would like to see. Even better is to arrange to meet the local inspector on site early in the job, so that the two of you can walk the job and the inspector can tell you what her/his typical concerns for the project might be. Remember that, although this may be your first office building, the building inspector has likely reviewed several and has experience worth mining.

The inspector will appreciate that you have sought his/her opinion, and will generally treat it as a mutual education and a "getting to know you" opportunity. The superintendent may wish to accompany you, which is his/her right – all the better, as it will foster mutual understanding. Harkening back to the opening quotation of this tale, the superintendent will also have the opportunity, with you as a witness, to ask the authority about inspection expectations.

RECORD the journey	Identify which reviews are required	**Summary Tools**	**Principles & Tools**
RESOLVE the issues	Independent i-A for each emergent issue	**i-WorkPlan (i-WP)**	Breakdown of review require-ments
REVIEW the results	Count & value each issue	**i-Actions (i-A)**	List of required reviews; review forms; non-conformance & deficiency forms
REMEMBER & learn to improve	Convert new knowledge i-A's into i-KH's	**i-KnowHow (i-KH)**	Amend i-WP and i-A form & template refinements

LINKS: To open the full list of Minimum Required Field Reviews, go to http://bit.ly/aagc14-1

## 15

Protecting The Public
Specialized Construction

"Yes, you have the right material, but your conduit can't penetrate it at a 45 degree angle – it must be perpendicular. Look at the rated construction." – Building Inspector, patiently, to Architect

What's the point? I had just had a bunch of firestopping penetrations rejected by the building inspector. The conduit penetrating the firestopping material was the maximum size allowed by the rated assembly. Unfortunately, that meant the penetration on an angle created an elliptical hole that exceeded the maximum permitted circular penetration. A subtlety that required considerable rework to make good.

3.3	INSPECTION
A.	Notify the Consultant when ready for inspection and prior to concealing or enclosing firestopping materials and service penetration assemblies.
B.	Unless otherwise stated in writing by all of the Consultant, local authority having jurisdiction and any building code consultant engaged on the project, do not enclose or otherwise render incapable of inspection any firestop installation until these parties have reviewed and accepted the installation.
1.	The Consultant will advise where the project is in a jurisdiction where the authority having jurisdiction specifically declines to provide inspection. In these cases, rely upon the Consultant's advice and do not await acceptance by the authority having jurisdiction prior to proceeding with work.

Figure 60 - My Specifications for Firestop Inspection

For smoke seals and firestops, the wide variety of construction assemblies and things of varying materials, shapes and sizes penetrating them has given rise to a large number of products and systems that have been tested as meeting specific standards as smoke or fire barriers. Architects and code consultants often identify the simplest of these on their drawings and in their specifications, but except for the simplest, smallest buildings they will almost inevitably miss some combinations. They must also be aware that certain firestopping materials may be incompatible with penetrating materials. There was an occurrence in my home town several years ago where a new smoke sealant turned out to dissolve a new plastic sprinkler piping – nasty while it lasted!

1.3	SUBMITTALS
A.	SAMPLES
1.	Submit samples in accordance with Section 01 3 00 - Submittal Procedures.
B.	PRODUCT DATA
1.	Submit product data in accordance with Section 01 3 00 - Submittal Procedures.
2.	Submit manufacturer's product data for materials and prefabricated devices, providing descriptions are sufficient for identification at job site. Include manufacturer's printed instructions for installation.
3.	Product Data and product certificates signed by manufacturer certifying that products furnished comply with requirements.
4.	Specifically identify any materials that are incompatible with the proposed firestopping materials.

Figure 61 - My Specifications for Firestop Submittals

2.1	**MATERIALS**	
A.	Fire stopping and smoke seal systems: in accordance with ULC-S115.	
	1.	Asbestos-free materials and systems capable of maintaining an effective barrier against flame, smoke and gases in compliance with requirements of ULC-S115 and not to exceed opening sizes for which they are intended and conforming to any special requirements specified in 3.5.
	2.	Firestop system ratings to meet requirements of the following fire separations:
	a.	Generally 2 hours horizontally and vertically through all concrete elements.
	b.	1 hour horizontally for elements penetrating between residential suites.
	c.	1 hour horizontally for elements penetrating between office suites.
	d.	1 hour vertically for elements penetrating between residential floors.
	e.	On floors with residential suites, 1 hour horizontally between suites and public corridors.
	f.	On floors with residential suites, 1 hour horizontally between suites and electrical closets, exit and elevator shafts and between public corridors and exit and elevator shafts.
	g.	2 hours horizontally between vertical shafts and adjacent occupancies, including between elevators and elevator machine rooms.
	h.	2 hours horizontally for conditions not noted above.

Figure 62 - My Partial Specifications for Firestopping Materials

What are the Principles & Best Practices: This complexity of solutions has given rise to subcontractors specializing in smoke and fire barrier systems. Many contractors now remove this aspect of construction from the various trades who see them on drawings and in specifications, instead asking for pricing that includes all situations.

Generally, designers should applaud and support this approach, as it provides a "one stop shop" for all matters of smoke and fire. By listening carefully to these subcontractors, designers can also learn much about the practical aspects of life safety systems, which will inform their future work and help them create better documents for situations where the firestopping subcontractor does not exist.

To be clear, unless the designers' documents specifically require a separate specialist subcontractor, there may not be one on the project. Historically, consultants have shied away from telling bidders how to organize trades to build a building – this may be a reasonable exception to consider. My compromise is to have a separate firestopping specification, recognizing this may make it easier but not mandatory to hire one specialist subcontractor.

Regardless how bid documents are structured, they should call for submittals covering all such products, and these should be carefully reviewed relative to building code requirements. Pay special attention to limitations such as the material and number of penetrating wires, pipes, etc., or the maximum size of hole that the system can be used in, then be prepared to review that on site – subcontractors, especially plumbing/mechanical, often cut much larger holes than they need for ease of running pipes and ducts. Oversized penetrations can usually be sleeved or modified to reduce the effective size of the penetration, but that will not happen unless/until a reviewer rejects work means and methods. Occupants' lives are at risk, so this is not an area to be relaxed about.

Another common challenge arises from confusion between firestop and smoke seal. Some materials do both jobs, but frequently rated "assemblies" are just that, a smoke sealant that often looks like caulking sitting atop horizontal construction or one or both sides of vertical construction. Usually both jobs need to be done.

By the way, my solution to the failed penetrations noted at the start of this tale was to frame out fire-rated boxes perpendicular to the angled penetrations, so that the maximum penetration size was not exceeded – messy and fortunately hidden in the ceiling space, but effective.

RECORD the journey	Specify penetration requirements	Summary Tools	Principles &
RESOLVE the issues		i-WorkPlan (i-WP)	Procedure for full inspection
REVIEW the results	Review every firestop penetration	i-Actions (i-A)	Submittal reviews; Field review; non-conformance or deficiency
REMEMBER & learn to improve	Add new approved materials as they emerge	i-KnowHow (i-KH)	

LINKS: To access the ULC Online Directory and find details of ULC assemblies, go to http://bit.ly/aagc15-1 to conduct your search (as of writing of AAGC, this was a free services).

## 16

Collaborating with the Superintendent

"No Consultant has ever before asked how we were actually going to build the build-ing, and how could we help each other." – Superintendent to Architect

What's the point? Design and construction is a collective effort. The owner-built home may be an exception, but even in that extreme case there are usually suppliers and installers of specialized elements.

A senior management colleague in the construction company where I currently work confided in me some time ago – *"Brian, come to the superintendent meetings if you want to see what's really going on."* I have since made it a habit of attending such meetings when I am visiting a site. Being wide awake in a meeting at 6am so that the superintendents can manage themselves and be on site before the subcontractors and workers arrive is no mean feat and gets my respect. Strong coffee also helps.

Architects are not generally in a position to attend such meetings but that does not prevent them from developing a good working relationship with the superintendent(s) of the projects they have designed.

What are the principles & best practices? On sites where there are multiple super-intendents, they will caucus regularly, often daily, usually before the work day starts and sometimes after as well. In addition, they will have regular meetings with the assembled subcontractors as well as issue-based meetings with individual subcon-tractors, usually around pre-mobilization, performance or scheduling issues. They will also attend regular OAC meetings (Owner-Architect-Contractor). When they are not in meetings they are generally on site, checking up on the status of work, identifying and managing conflicts, bottlenecks, etc., or walking around with Consultants, Owners and regulatory agencies. They also prepare and update reports and schedules and generally keep a daily diary that may also be a report of activities to "head office." Somewhere in there they eat and sleep.

Contractors will deem this self evident, but believe me when I say that many other project team players, including some clients and consultants, will not realize that all work on site flows through the superintendent(s). Superintendents may delegate if they have juniors to delegate to, but they are responsible for everything that happens on the construction site. These folks are busy – we can make their lives easier or more difficult. Generally when their lives are easier, so are those of consultants and clients. As the quote at the start of this tale suggests, I have made it a habit for years to meet with the superintendent at the beginning of my involvement as consultant on any con-struction site. I will start by asking the superintendent to explain the basic construction sequencing in her/his own words, regardless whether there is a published schedule (that I may or may not be able to understand). For example, in an L-shaped building some superintendents will organize the work to progress up the building one floor at a time, i.e., Level 2 East, Level 2 West, Level 3 East, Level 3 West, etc. Other super-intendents will race up one wing before starting the other. Whatever the sequence, the chances are that I can adjust my rhythm of field services to match the superinten-dent's.

Why is this important? If I visit a portion of the building that is less advanced, I may list a large number of "deficiencies" that the superintendent sees as work in progress/incomplete work. If this misunderstanding continues long enough, it will grow into mistrust and even dislike. And the paperwork will multiply.

Conversely, the contractor's field staff needs to understand that consultant agreements do not require them to follow the contractor's schedule. and they do not provide for the consultants to be at the contractor's beck and call! I have on several occasions politely informed my client that my time was being abused by the contractor and I would start charging hourly for what are in effect additional services. That usually precedes a conversation between client and contractor that irons the issues out. Occasionally the client will call back and approve the additional services based upon scheduling or marketing requirements that are beyond my control. Either way I am covered.

Just as many designers do not understand what happens on a site, so too many superintendents (and project managers for that matter) do not understand how designers work. Sometimes it is just as difficult and time consuming to draw a complex detail as it is to marshal the subcontractors, suppliers and installers who will build it. Sometimes it is just as time consuming to properly answer a Request for Information (RFI) as it is to execute the answer. It's not necessary to teach superintendents what happens in design offices, just that every design office is different and outside their experience. You cannot assume that any superintendent has a deep knowledge of how designers work.

There is only one solution to working with the superintendent and that is to effectively communicate with the site. By all means try on your standard procedures, reports and forms, but also listen carefully to the reaction and be prepared to adapt. The tales that follow are some examples of listening and adapting.

RECORD the journey	Confer about scheduling	Summary Principles & Tools	
RESOLVE the issues	Confer about the issue resolution	i-WorkPlan (i-WP)	
REVIEW the results	Meet with the superintendent each time on site	i-Actions (i-A)	Consider the superintendent's form
REMEMBER & learn to improve		i-KnowHow (i-KH)	

## 17

Administering
Whose Contract Is It Anyway?

"My contract is with the Owner, so I represent the Owner's interests during construction" – Architect to Contractor

What's the point? All client/consultant agreements require that the consultant perform magic during construction ("What's taken you so long to figure this out," you are thinking). While the client retains the consultants to design the project, as soon as construction starts the consultants are expected to go through a metamorphosis like moths to butterflies and suddenly become independent arbiters of the construction contract between the client and contractor. This means they are expected to decide any issues impartially as between the client and contractor. This is particularly difficult, given the client is still paying the consultants' fees, and even more tricky when an issue reveals a shortcoming in your documents. But that is the expectation.

What are the principles & best practices? Conventional construction contracts will mention the Consultant in the contract, even though the Consultant is not a signatory to it. The standard Canadian construction contract mentions "the Consultant" on 27 of its 33 pages, a total of 183 times, not including the table of contents and headings. If you ignore the responsibilities inherent in those +/-183 entries you will damage your reputation.

GC 2.1 AUTHORITY OF THE CONSULTANT

2.1.1   The *Consultant* will have authority to act on behalf of the *Owner* only to the extent provided in the *Contract Documents*, unless otherwise modified by written agreement as provided in paragraph 2.1.2.

2.1.2   The duties, responsibilities and limitations of authority of the *Consultant* as set forth in the *Contract Documents* shall be modified or extended only with the written consent of the *Owner*, the *Contractor* and the *Consultant*.

2.1.3   If the *Consultant's* employment is terminated, the *Owner* shall immediately appoint or reappoint a *Consultant* against whom the *Contractor* makes no reasonable objection and whose status under the *Contract Documents* shall be that of the former *Consultant*.

GC 2.2 ROLE OF THE CONSULTANT

2.2.1   The *Consultant* will provide administration of the *Contract* as described in the *Contract Documents*.

2.2.2   The *Consultant* will visit the *Place of the Work* at intervals appropriate to the progress of construction to become familiar with the progress and quality of the work and to determine if the *Work* is proceeding in general conformity with the *Contract Documents*.

2.2.3   If the *Owner* and the *Consultant* agree, the *Consultant* will provide at the *Place of the Work*, one or more project representatives to assist in carrying out the *Consultant's* responsibilities. The duties, responsibilities and limitations of authority of such project representatives shall be as set forth in writing to the *Contractor*.

2.2.4   The *Consultant* will promptly inform the *Owner* of the date of receipt of the *Contractor's* applications for payment as provided in paragraph 5.3.1.1 of GC 5.3 – PROGRESS PAYMENT.

2.2.5   Based on the *Consultant's* observations and evaluation of the *Contractor's* applications for payment, the *Consultant* will determine the amounts owing to the *Contractor* under the *Contract* and will issue certificates for payment as provided in Article A-5 of the Agreement - PAYMENT, GC 5.3 - PROGRESS PAYMENT and GC 5.7 - FINAL PAYMENT.

2.2.6   The *Consultant* will not be responsible for and will not have control, charge or supervision of construction means, methods, techniques, sequences, or procedures, or for safety precautions and programs required in connection with the *Work* in accordance with the applicable construction safety legislation, other regulations or general construction practice. The *Consultant* will not be responsible for the *Contractor's* failure to carry out the *Work* in accordance with the *Contract Documents*. The *Consultant* will not have control over, charge of or be responsible for the acts or omissions of the *Contractor*, *Subcontractors*, *Suppliers*, or their agents, employees, or any other persons performing portions of the *Work*.

Figure 63 - Part of 1 of 33 pages of a Construction Contract

There is no magic to interpreting the construction contract – as another colleague says, *"Read the words, then do what the words say."* Until you are more seasoned, you should always read the construction contract. Try what I did and search where the word "Consultant" occurs in the agreement - that's your scope. As you become more knowledgeable about contracts, never forget to search out the supplementary conditions and other amendments to the "standard" agreement.

Don't make the mistake I once did and assume that because the client is the development arm of an insurance company, the agreement will therefore be balanced. The worst contract I have ever seen was the "standard" of a large insurance company. I was so aghast that I sent it to my liability insurer who, at my urging, wrote such a scathing email review of the agreement that I was able to supplant it with a real "standard" agreement. By the way, that's the level of support service you should expect to receive from your insurer.

A colleague tells me of a senior lawyer in a national firm who told him he was particularly proud of his contracts, not because they had helped his clients win in court, rather because they had never been to court! His contracts were balanced and fair to both parties to the agreement and were accepted as such.

If you are armed with an understanding of the construction contract for your project, you will begin to recognize the origins of many questions and comments from the superintendent. If you don't understand where a query is coming from, remember there are no dumb questions and ask for an explanation. If you have made an effort to work with the superintendent, he/she will look at it as a teaching opportunity and if nothing else, you will learn how the superintendent thinks about one element of the design and construction process, never a bad thing.

Consultants who are known as fair interpreters of the construction contract will attract more qualified bidders to their projects (*"good to work with...for an architect"*) and better prices (*"fairly impartial...for an architect"*). They will spend less time interpreting documents (*"fair but firm...not much point arguing"*).

RECORD the journey	Confer about scheduling	Summary Principles & Tools	
RESOLVE the issues	Confer about the issue resolution	i-WorkPlan (i-WP)	
REVIEW the results	Meet with the superintendent each time on site	i-Actions (i-A)	Consider the superintendent's form

LINKS: In Canada, for standard short form client/architect agreements, start with the RAIC Document 7 at http://www.raic.org Professional Resources -> Contract Documents - note that several provinces have recommended variations. For larger projects, look at RAIC Document 6.

For American contracts of various kinds involving architects, go to http://bit.ly/aagc-16-aia - note, most of the documents at AIA are NOT free.

For British contracts of various kinds involving architects, I was completely flummoxed by the RIBA website. Maybe an enterprising reader can assist!

For Canadian construction contracts of various kinds, go to the CCDC site at http://bit.ly/aagc-16-ccdc - NOTE, most of the documents at CCDC are NOT themselves free, but much CCDC information is free. CCDC documents are consensus-based amongst designers, builders, lawyers and regulators, hence are widely used in Canada.

For British construction contracts of various kinds, go to http://bit.ly/aagc16-jct where JCT maintains a useful site to a myriad of contracts!

## 18

Administering
Who Pays the "Extra?"

"Our excavator dug the road bed about a foot deeper than the drawings show because the bearing was not good enough. So according to the unit rates in our contract, the Owner owes us $$$" – Contractor to new Architect

What's the point? I was at my first ever project monthly claim review meeting – not my first project but the first where the principal of the firm deemed me competent enough to review the contractor's claim (before the principal checked it). There wasn't much to look at, some excavation and site works, so only the civil engineering consultant was with me when we were presented with the quotation above. The "extra" money claimed was significant. I was dumbstruck.

Fortunately the civil consultant was experienced, not dumbstruck. "This is the first I've heard of this", he said. "You did not call me to verify the site conditions before you dug and backfilled – there is no claim for extra." That was the polite version of what he said.

What are the principles & best practices? Because claims for extra can be contentious, construction contracts are usually very clear about them. Where a unit rate is included in a contract, it is usually explicit (and was in this case, I just forgot): the Consultant must be called to site to verify the conditions before extra work associated with the unit rate is undertaken. Without verification by the person(s) certifying payment, there is generally no basis for a claim for extra.

Another fairly typical example of this is when the contractor uncovers boulders during excavation. Most bid documents will identify that rock in excess of a certain volume will be difficult to remove, was not expected and merits an extra charge based on unit rates. Upon discovering rock, a savvy contractor will inform the geotechnical engineer (who needs to advise the prime consultant). The geotech will visit the site and verify the rock volume, the contractor will remove it (may involve blasting) and a claim for extra will follow.

RECORD the journey	Specify requirements for consultant review before proceeding with "extra"work	Summary Principles & Tools
RESOLVE the issues	Specify unit rates where possible to easily quantify extras	i-WorkPlan (i-WP)
REVIEW the results	Have the responsible consultants review	i-Actions (i-A)

LINKS: To look at a sample excavation control log, go to http://bit.ly/aagc18-1

## 19

Administering
The Dreaded RFI

"I feel like the contractor is picking apart my design, looking for extra's"
– Architect to colleague

"So many construction document sets are not sets at all – they're incomplete and poorly coordinated. I need some way to figure out how to build the building!"
– Contractor to colleague

What's the point? Requests for Information (RFI's) arise from contractors. They occur frequently in construction, yet are frequently misunderstood. Many consultants believe they foster laziness among contractors, who don't need to bother to learn what the construction documents say. Many contractors believe they foster laziness among consultants, who don't need to bother to complete or coordinate their documents before issuing them for construction. There is some truth to both perspectives.

What are the principles & best practices? When bid documents are subsequently issued for construction (IFC), the contractor should be able to build from them – it's that simple. When, after reviewing the IFC documents, a contractor (including subcontractors and suppliers) cannot determine how to supply or install an element of construction, a RFI is issued from the contractor through the architect/ prime consultant to the designer(s) who can answer the query. Just like a contractor assembles pricing for a contemplated change order, so, too, for a contractor's question, the architect assembles answers to RFI's and communicates them back to the contractor, and work carries on.

The figure below illustrates an i-A/RFI. Notice that below the i-A "tags" there are only five items to fill in, which capture the essence of an RFI:

Item:	RFI 02 - Request for Information			Action Number:	A00045 R0	Priority:	04 by next meeting or Due Date
Type:	Sample			Subtype:		Element:	07 Exterior walls general
Status:	Open			Due:	2014/05/20	Opened: 2014/02/12	Closed:
Identifier:	Level 3 Building 4			Originator:			
	RFI 02						
Question/Confirmation:	How can we firestop the juncture of the two walls meeting at an angle between Suites 304 and 305?						
Requested by:	Superintendent Bob						
Answer:	Refer to spec section 07900 - Firestopping, Section 2 - Materials. Use Firestopping system 4, which should accommodate the odd-shaped penetrations in this area.						
Answered by:	Referred By						
Notice of Change to follow?	No						
Controlled document	Rev 00 - 101203						

Figure 64- Structure of an RFI

The five essential RFI components are:

1. What is the question itself? – Contractors quickly develop expertise in asking clear RFI questions, because unclear questions cause delay and poor answers. Sometimes the question will come with a suggested answer in the form of a recommendation. If there is no cost to the recommendation, this is a good start.

A small minority of contractors sees RFI's as an opportunity to seek out extras. You can usually tell when this is the approach because you will generally be inundated with a large number of RFI's, and your responses will be debated – *"That detail is not similar to the drawn detail" "We did not allow for that in our bid because it's not drawn",* etc. The second *"We did not allow for that..."* quote above is interesting, since the contractor has submitted a price to construct the entire building. Even if you have completely missed a detail, or the "similar to" detail is not really similar at all, so long as that corner window or unusually placed skylight is shown on the drawings, then the contractor had to make an allowance for it. In the worst case scenario (for cost), the extra cost should be limited to perhaps a couple of strips of peel and stick.

The vast majority of contractors just want an answer to their question, and to get on with the job.

2. Who is asking the question? – Receiving a large number of RFI's, especially at the start of a project, is not necessarily an indication that the contractor is looking for extra's. First estimators (during bidding), then project managers (during project startup), then superintendents (during mobilization) are all expected to become knowledgeable about the construction documents. Each will develop a list of items unclear to them. Some of the estimator's questions will be asked and answered during tendering, but some estimator assumptions will be passed along to the project manager, who has ultimate responsibility for project completion "on time and on budget." The project manager will therefore likely ask questions about elements that the estimator made assumptions about, often during the "buy out" period when subcontractor work scopes are being finalized. Finally, the superintendent is responsible for the means and methods to get the building built on time and on budget, so will ask experience-based questions, especially about design elements that may be challenging (or occasionally impossible!) to construct as designed. We generally call these "constructability" questions.

3. What's the answer? – Before answering a RFI, the respondent needs to consider the broader design and construction implications. It is not the contractor's job to identify building code, building envelope or "green" building implications of a response, and certainly not design implications. The builder has not been part of the design process, so may ask a question that misses the mark because of that lack of context.

RFI's can lead to claims for extra if they highlight weaknesses in Issued for Construction (IFC) documents. Therefore, avoid answers couched in phrases like *"The detail should have shown...", "I meant to include..."* Better to stick to the facts and use the active imperative of specifications: *"Add this", "Delete that", "Move this from here to there",* etc. If IFC documents are weak, you can pretty much guarantee that RFI's will lead to requests for change (i.e., more money or time, or both).

Often the answer is a reference to a detail or specification that may not have been as obvious as it should have been. There is nothing wrong with this. Every human being is different and the mental logic paths we use vary widely. Be as patient with the questioner as you would expect them to be with you.

The answer to an RFI should never be abrupt or demeaning, although it can be active imperative like a specification: "Go to this location in the documents. Look here for the answer", or "The answer is ..."

A RFI answer should never be unclear or subjective. *"The design intention is…", "It's implied in the specs that…"* are never acceptable answers. You might just as well say "BFO." (Builder Figure Out). Alternatively, the kindergarten jingle applies: *"You get what you get and you don't get upset."*

4. Who's answering the question? – It is good practice to have a peer or principal review the draft of RFI responses for two reasons: 1). Does the answer seem logical and fully answer the question asked? You want the process to be crisp, clear and not drawn out. 2). Does the answer expose the firm to any potential claims for extra? Regrettably, the answer is sometimes yes, but this should be the exception.

5. Is the result of the answered question a change to the work? – This RFI element would generally be filled in by the contractor, reacting to the RFI response. Where this occurs it needs to be treated just as any claim for extra, i.e., evaluated on its own merits, in the context of the contractor's commitment to build a complete project for a price.

What is the most efficient solution to manage an RFI? What are the best tools? As was noted at the outset, I use an i-Action (i-A) with a specific layout to manage an RFI. When I am wearing the designer's hat and receive an RFI, I transpose the question into an i-A because this is the most efficient manner for me to manage it. The originating document is usually an email or an attached PDF document; either way it is usually easy to copy and paste the RFI into the i-A.

Why bother? I have attended many meetings where RFI's consumed vast amounts of time, largely consisting of statements like *"Oh, I didn't get the attached drawing, can you resend?"*, or *"I'm sure I forwarded the question to the mechanical engineer. Oh, he's not here! I'll chase him…"*

Delayed response to RFI's can result in claims for delay from an astute builder, so they need to be treated seriously. An i-A puts you in control. Its start can be dated, due dates (perhaps your reasonable ones, not others' unreasonable ones) can be identified, which will result in an automatic reminder to the assignee and assignor. Each time you respond to or refer an RFI, in fact every element of the discussion, is automatically aggregated to the originating i-A – the workflow is fully managed. At any moment or meeting you can assemble the list of open or closed RFI's, RFI's assigned to certain individuals in the workflow, RFI due dates in order, all sorts of good stuff including all of the subsequent discussions!

To optimize using i-A's for RFI's: use the Identifier to pinpoint a location; set up a due date for each step in the workflow, so that reminders are issued before due dates, including where the RFI is referred to another consultant; use the Element field to select the spec section or general area of construction; use the Originator field to identify individuals generating RFI's:

Action								
Now Responsible:			Referred By:					
Item:	RFI 02 - Request for Information		Action Number:	A00045 R0	Priority:	04 by next meeting or Due Date		
Type:	Sample		Subtype:		Element:	07 Exterior walls general		
Status:	Open		Due:	2014/05/20	Opened:	2014/02/12	Closed:	
Identifier:	Level 3 Building 4		Originator:					

RFI#	02
Question/Confirmation:	How can we firestop the juncture of the two walls meeting at an angle between Suites 304 and 305?
Requested by:	Superintendent Bob
Answer:	Refer to spec section 07900 - Firestopping, Section 2 - Materials. Use Firestopping system 4, which should accommodate the odd-shaped penetrations in this area.
Answered by:	Referred By
Notice of Change to follow?	No
Controlled document	Rev 00 - 101203

Figure 65- Sample RFI in i-A format

As you can see from the example above, the total document remains succinct, but can be searched by any of the criteria between the horizontal bars, or any combination of them.

RECORD the journey	Capture each RFI	Summary Principles & Tools	
RESOLVE the issues	Track each RFI to resolution	i-WorkPlan (i-WP)	Specific procedures
REVIEW the results	Identify if an RFI leads to a change	i-Actions (i-A)	Specific forms or templates
REMEMBER & learn to improve	Feedback RFI's into the knowledge database, specs & details	i-KnowHow (i-KH)	Knowledge items to look for

LINKS: To access the sample RFI form, go to http://bit.ly/aagc19-1 and click on the blue http file name at the bottom left of the screen.

20

Contructing
Complete & Sequential Detailing

"Half the details are missing"– Building Envelope Consultant to Architect

"You try building that the way it's been drawn – impossible!"– Superintendent to Project Manager

"I can't review the details that are not even drawn!"– Building Envelope Consultant to Architect

What's the point? At the architectural firm where I learned most about communicating through construction drawings, we were taught to cut multiple 1:50 (1/4"=1'-0") sections through a building design, on the theory that challenges and conflicts could not be hidden at that scale, would then need to be detailed. Notwithstanding to-day's CADD and BIM environments, this thought process has continued to inform my work, including specialized building envelope work where I review the designs of other architects. The cover of this book is a portion of a schematic I developed to identify all of the typical sectional and plan details I should expect to find in construction documents I was reviewing – as per the third quotation above, you can't review what you don't see.

As I work through a document set, I tick off the details as I find them and cross out those that do not apply – a graphic checklist, if you will. I also use the graphic to identify to designers a list of missing details. The checklist was developed around wood frame construction, but the details apply equally to steel and concrete structures:

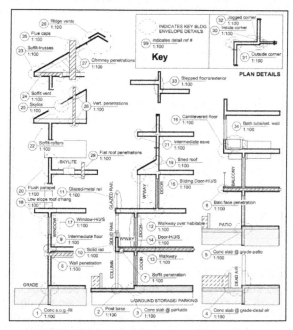

Figure 66- 35 Key Building Envelope Details

The second quote above from a superintendent concerned about constructability (a term I had never heard before working for a contractor) highlights an area of significant confusion between designers and builders. Virtually all construction documents assert that the designer is not responsible for the contractor's sequencing of trades, means and methods, etc. This is not unreasonable, as this is the contractor's expertise.

However, some designers take this to the extreme by creating details that are not conventionally constructable as drawn – typically the conventional sequence of construction does not work.

What are the principles & best practices? What do I mean by "the conventional sequence of construction"? Typically, a hole is dug, concrete is poured for foundations, and structural superstructure is erected. The building envelope cladding, including roofing, is affixed to the cladding. When the interior is weather protected, services are roughed in, then interior construction such as partitions is erected. Finally the finishes are added.

To illustrate the importance of sequencing, the exterior cladding of the first building I ever designed included an exterior rainscreen cavity wall – an exterior brick veneer and interior painted concrete block (the building was for equipment, not people). Sandwiched between the two block layers was a layer of rigid insulation and an air space. (This is long enough ago that I will not apologize for the many cold joints and other energy discontinuities.) I was working at a summer job for a large utility company, and my boss' attitude to design detailing was "We have a room full of architects and draftsmen – they can help you figure it out." And they were very helpful and patient. After much looking over of shoulders and agonizing, I brought my typical wall section to a particularly helpful architect:

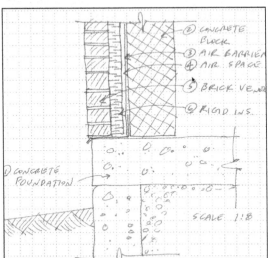

Figure 67- Mis-designed Wall

He could not resist a smile but was otherwise patient with me. "Can you describe how you would build this wall?" I recited the numbered sequence in the detail above. After I finished with #6, he asked, "And how will you fasten the rigid insulation to the inside face of the brick?" Of course, I had no answer. Not only had I not understood the basic concept of a rainscreen, I had placed the rigid insulation in an impossible location – if the builder built from the inside out (the usual way), there was no way to affix the insulation to the brick veneer. And if the builder constructed from the outside in (an unusual sequence except for a few structures like barns), there would then be no place to affix the air barrier.

Of course, we resolved the error by simply moving the insulation to the other side of the cavity. Disaster averted and lesson learned!

Any time that the design of a project detail interrupts the typical flow of construction, it is incumbent on the designer to consider this in detailing. Another quick and more recent example: I worked on a project with large (60m/190 feet long) internal metal soffits designed as single segments with no joints. The sequencing of work involved these being erected after the curtain wall was mostly in place. This would not work, could never work. The issue was eventually resolved (the soffits were segmented, which was also important to manage their expansion and contraction), but not without some argument and bad feeling between designer and builder.

Although the soffit was an extreme example, I have found it greatly helps my work to develop details in typical construction order. As a simple example:

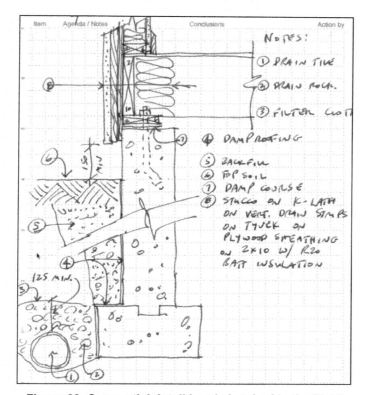

Figure 68- Sequential detail hand sketched in the Field

Here I have developed the detail and numbered the detail notes in typical construction sequence without any guarantee on my part that a). The builder will construct in that order, or b). The builder may have other construction challenges. That's not actually the point. The point is that I have thought through the detail in a sequence that makes construction sense and is buildable as such.

Even with my years of experience, I periodically sketch a sequential detail only to discover the sequencing is impossible, or at least challenging. That detail never sees the light of day.

Incidentally, superintendents love sequential details even if they ultimately build in a slightly different order. The approach tells them you are thinking about their challenges and helps them with their challenges with trades. Increasingly, some construction workers lack even rudimentary education about construction – sequential details can help the superintendent focus workers on what to do in what order.

RECORD the journey	Use sequential detailing to create logical details	Summary Principles & Tools	
RESOLVE the issues	Use peer review to identify issues	i-WorkPlan (i-WP)	Sequential detailing procedure
REVIEW the results	Ensure all indicated details are provided	i-Actions (i-A)	
REMEMBER & learn to improve	Standard sequential details	i-KnowHow (i-KH)	

## 21

Demountable Detailing

"So you're telling me that when the windows need replacement in 25 years the owners get to tear out the 50-year brick wall to get at them?" – Project Architect to Principal Architect

What's the point? At the risk of straying too far into design, the quote above was directed at me by an employee detailing a brick clad wood-frame residential building for a local developer. I had been trumpeting the importance of "green" building design and she caught me out!

Specifically, I was not paying attention to the realities of flanged windows installed in openings where the brick returns to the window edges:

Figure 69- Break the Brick to Remove the Window

When the window above needs replacement, its removal will inevitably damage the surrounding brickwork due to the 1" (25mm) flange used to nail the window frame to the framing behind.

What are the principles & best practices? Embarrassed by my employee's realization, I directed her to figure out how to detail flanged windows to be replaceable without damaging masonry:

Figure 70- Removable Flanged Window

The simple solution: Detail a trim piece around the window. You can't see it from the photo, but the trim piece is actually screwed in place with grommetted stainless steel screws, so that the trim as well as the window can be removed without damage. And of course the screws underline the demountability, hence sustainability, of the result.

RECORD the journey	Consider the life and durability of material sequences	**Summary Tools**	**Principles &**
RESOLVE the issues	Use construction sequencing logic to identify issues	**i-WorkPlan (i-WP)**	Include a durability plan in all projects
REVIEW the results	Use a durability plan to verify a sustainable long term plan	**i-Actions (i-A)**	Use to track issues emerging during design
REMEMBER & learn to improve		**i-KnowHow (i-KH)**	Record expected maintenance and replacement intervals for materials & systems.

LINKS: For a sample of a durability plan developed by my former employers Morrison Hershfield Ltd. from Canadian standards, refer to http://bit.ly/aagc21-1

## 22

Scheduling -
Whose Governs

"You have to provide your services to meet our construction schedule, otherwise we will never finish on time or budget." – Contractor to Consultants

What's the point? The construction schedule is one of the most fundamental controllers of the contractor's efforts. It is no wonder the contractor may assume the consultants will work to their schedule, but this is not exactly the case. First, in most cases there was no schedule in place when the consultants were hired, beyond perhaps "anticipated completion autumn of 2016." Secondly, while the contractor controls work on the site, including subcontractors, suppliers and installers, this control does not include consultants, with whom the contractor has no contract.

What are the principles & best practices? So what is the happy medium? The consultants will usually receive a copy of the construction schedule early in the course of the project construction, often at or before the first site meeting. Sometimes the schedule is a simple bar chart; more often in the projects of a scale that many architects are engaged in the schedule will be a Gantt chart, that is a bar chart with accurate start and end dates and including the interrelationships between activities ("...this can't start until that has ended, or not until that has progressed for six days...")

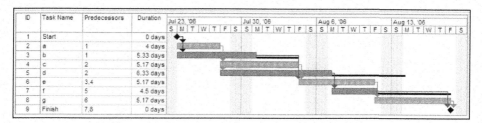

Figure 71- Simple Gantt Chart[1]

To be useful for consultants, a good schedule should include: when various submittals such as shop drawings are to be sent to the consultants, and when they are due back; the timing of tests and mockups; when portions of the project will be ready for consultant field review; a list of long order items and their associated submittal review dates; a list of staged or phased tenders, such as many fast track or construction management projects; etc. It is the contractor's responsibility to schedule events that are a consequence of long order times or staged tenders – they control these, not the consultants.

If any of this detail is missing, ask for it as soon as you have reviewed the schedule, and if necessary raise it as an issue at the next site meeting. Why? To make it clear that you have read and understood the schedule and will work with it, but only if it works with you.

---

1   Wikipedia, http://en.wikipedia.org/wiki/Gantt_chart

Your architectural association will probably tell you that you have approximately ten working days to review most submittals, but the reality is that submittals are sometimes late, with no change to the expected return date. You may wish to indulge the contractor a few times around accelerated review times for late submittals, but if this becomes a habit it generally means the contractor is not managing the site, subcontractors and suppliers well, which is not your responsibility and should not place upon you any time penalties.

One way to manage your time around the contractor's schedule is to extract from the schedule those dates that pertain to your in-office (submittals) and on-site (mockups and field reviews) time. I have a master list of submittals for a project, which I customize for each project's details:

List of Required Submittals   logo here

Date: _____

Project # _____

Project Name: _____

'+'= req'd, '√'=completed, r' = rejected

NOTE: Each submittal will be considered via a project specific submittal review.

Spec division/ item	Mockup req'd	Manu. Lit./ Document	Shop drawings	Test data	Sample	Warranty duration (minimum years)
01 Construction schedule		+				
01 Proof of insurance		+				
01 Bonds		+				
01 Proof of current insurability by provincial workers compensation authority		+				
01 Site superintendent resume		+				
01 Proof of construction safety certification of onsite personnel [1]		+				
01 Fire safety Plan [2]		+				
02 Drain Mat @ parkade walls [3]	+	+				
02 Drain Mat @ suspended slab [4]	+	+				
02 Hard landscaping elements		+			+	
02 Soft landscaping		+				
03 Concrete mix design		+		+		
03 Concrete waterstops	+	+				
03 Precast concrete			+	+	+	
03 Grout		+				
04 Clay masonry units	+	+	+	+	+	

Figure 72- Partial Submittal List

I can take this list and extract from it scheduled dates to replace the "x"'s that indicate things I expect to see:

List of Required Submittals
logo here

Date: _____

Project # _____

Project Name: _____

'+'= req'd, '√'=completed, 'r' = rejected

NOTE: Each submittal will be considered via a project specific submittal review.

Spec division/ item	Mockup req'd	Manu. Lit./ Document	Shop drawings	Test data	Sample	Warranty duration (minimum years)
01 Construction schedule		+				
01 Proof of insurance		+				
01 Bonds		+				
01 Proof of current insurability by provincial workers compensation authority		+				
01 Site superintendent resume		+				
01 Proof of construction safety certification of onsite personnel [1]		+				
01 Fire safety Plan[2]		+				
02 Drain Mat @ parkade walls[3]	141029	141025				
02 Drain Mat @ suspended slab[4]	141029	141025				
02 Hard landscaping elements		150625			150625	
02 Soft landscaping		150625				
03 Concrete mix design		140924		140924		
03 Concrete waterstops	141104	141101				
03 Precast concrete			+	+	+	
03 Grout		+				
04 Clay masonry units	+	+	+	+	+	

Figure 73- Submittal list with Dates filled in

I generally bring this list to each site meeting and use it to manage my time going forward. When it is evident to the contractor that you are managing your time in a reasonable and responsive way, there will be an increased likelihood the contractor's time will be similarly managed. Also, where dates are changed, you can enter the new dates and reissue what is, in effect, a list of the things you need to see and participate in. If you feel extra fees are required because of changes, you will have an explicit record of what changed, and when.

Each submittal can also be handled as an i-A, which allows items to be considered individually, in filtered groups or all together.

Identifier	Project #	Number	Item	Floor	Due Date	Opened	Closed	Status	$Value	Priority	Type - Subtype	Company - Now Responsible	Company - Referred By	Originator	Element
07200-3	131218	A00129 R0	Waterproofing membrane		2014/10/16	2014/09/23		Open		11 Reference	Submittal - for review	Quality-by-Design Software Ltd. - Brian Palmquist	Quality-by-Design Software Ltd. - Brian Palmquist		07 Waterproofing
	copyright 2014 -	A00120 R0	Cladding		2014/10/21	2014/09/18		Open		11 Reference	Submittal - reviewed as noted	Ledcor Construction Limited - Brian Palmquist	Quality-by-Design Software Ltd. - Brian Palmquist		07 Exterior walls general

Figure 74- Individual Submittal i-A's

In this instance, changes to start and end dates can be added to the individual items. If you are using an i-WP approach, those changes can be diarized so that, if they arise from the contractor's project management, they do not give rise to any delay claims.

103

RECORD the journey	Create and review a schedule	Summary Principles & Tools	
RESOLVE the issues	Identify & schedule consultant work arising from the schedule	i-WorkPlan (i-WP)	Diarize schedule changes
REVIEW the results	Use the i-WP diary and/or i-A's or similar to manage each schedule element	i-Actions (i-A)	Unique i-A for each submittal & issue arising
REMEMBER & learn to improve		i-KnowHow (i-KH)	Capture new knowledge from i-A's

LINKS: To access my full submittals list and the location where I diarize submittal changes, go to http://bit.ly/aagc22-1. The list is the Word doc at the bottom left. To access the i-A for Submittal review, go to http://bit.ly/aagc22-2. The submittal review form is the blue http file name near the bottom left.

## 23

Scheduling & Tracking
Submittals with i-A's

What's the point? Each standard i-WorkPlan (i-WP) should have one or more procedures around submittals – collecting a list of submittals from the contractor, reviewing submittals, reviewing proposed alternatives and substitutions, etc. Each Procedure around submittals will have an i-A template for reviewing submittals:

SUBMITTAL REVIEW:

This form is used to transmit & evaluate a submittal, including a proposed alternative or substitution. To complete this form or reply to this email, select "Reply All", update information & select "Send". "B" = Builder, who indicates: "X" = details attached/ should be attached; "N/A" = not applicable on this occasion; "D" = Designer, who indicates "R" = Reviewed; "RN" = Reviewed as Noted; "RR" = Revise & Resubmit; "X" = Rejected. if you are using an iPhone or iPad you can select a "COMMENT/ SIGN OFFS" text box and dictate comments.

D	B	Type of submittal:	Spec'd	_Alternative _Substitution	Comments
	x	Indicate reason for proposed alternative or substitution:		_Cost savings _Schedule improvement _Quality improvement _Other (specify):	
		Description:	Original	Proposal	
	x	Specification section/clause:			
	x	Drawing reference:			
		Submittal information:			
		Manufacturer's literature			
		Installation instructions			
		MSDS data sheets			
		Sample warranty			
		Product sample			
		Shop drawings			
		Other (specify):			
		Review proposal for:			
	x	clearance dimensions between products and surrounding enclosure elements			
	x	"n.i.c." in someone else's contract.			
	x	any qualifications on the submittal			
		Other:			

Figure 75- Submittal Review Report Body

What are the principles & best practices? The detail included in a submittal need only be addressed when the submittal arrives. I simply duplicate the i-A template once per submittal, lodging each i-A with the appropriate submittal review procedure and naming it with the name of the submittal. So long as I have identified the Type of i-A as Submittal, I can thereafter call up the list of all Submittals to check status, and manage each one as to due dates, responsibilities, details, etc. as they arise.

Identifier	Project #	Number	Item	Floor	Due Date	Opened	Closed	Status	$Value	Priority	Type - Subtype
07200-3	131218	A00129 R0	Waterproofing membrane		2014/10/16	2014/09/23		Open		11 Reference	Submittal - for review
	copyright 2014 -	A00120 R0	Cladding		2014/10/21	2014/09/18		Open		11 Reference	Submittal - reviewed as noted
	131218	A00049 R0	00 Submittal Review Report			2014/02/12		Open		11 Reference	Submittal - for review
	140401	A00034 R0	00 Material Sample Review Report			2014/02/12		Open		11 Reference	Submittal - for review
Building 4	copyright 2015-	A00034 R0	00 Material Sample Review Report		2014/05/21	2014/02/12		Open		11 Reference	Submittal - for review

Figure 76- Submittal List i-A's

On my largest project to date we used this approach to manage more than 2900 separate submittals.

RECORD the journey	Capture each submittal	**Summary Principles & Tools**	
RESOLVE the issues	Capture review & related issues to the originating i-A	**i-WorkPlan (i-WP)**	Specific procedures
REVIEW the results	Use the i-WP diary and/or i-A's or similar to manage each schedule element	**i-Actions (i-A)**	a separate i-A for each submittal
REMEMBER & learn to improve		**i-KnowHow (i-KH)**	

LINKS: To review typical submittal procedures, go to http://bit.ly/aagc23-2

## 24

Paying
How Much do I owe You? - Payment Certification

"But I already paid 45% to the mason, you can't cut me back to 40%"
– Contractor to Payment Certifier

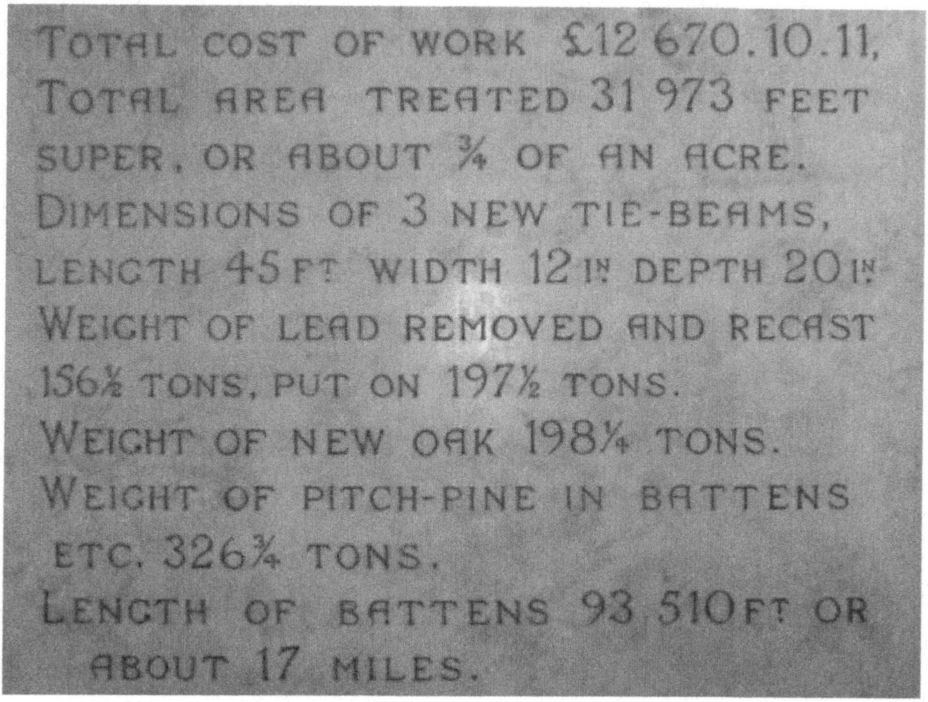

TOTAL COST OF WORK £12 670.10.11,
TOTAL AREA TREATED 31 973 FEET
SUPER, OR ABOUT ¾ OF AN ACRE.
DIMENSIONS OF 3 NEW TIE-BEAMS,
LENGTH 45 FT WIDTH 12 IN DEPTH 20 IN
WEIGHT OF LEAD REMOVED AND RECAST
156½ TONS, PUT ON 197½ TONS.
WEIGHT OF NEW OAK 198¼ TONS.
WEIGHT OF PITCH-PINE IN BATTENS
ETC. 326¾ TONS.
LENGTH OF BATTENS 93 510 FT OR
ABOUT 17 MILES.

Figure 77- This payment certificate is not editable (it's in stone!)

What's the point? Next to building envelope failures and litigation arising from bidding problems, payment certification is the architect's major source of claims. Why? Because the amount of money involved can be huge and there are so many ways to make a mistake.

What are the principles? Traditionally, architects manage a process each month during a project where the contractor provides a detailed invoice for the work completed that month, based on a previously agreed invoice template called a "schedule of values." The architect and other consultants review the monthly invoice under the architect's overall management, and agreement is reached about how much the contractor should be paid for the previous month's efforts. Sometimes that recommended amount is different from the contractor's initial invoice, and the process by which the initial invoice is vetted can give rise to significant discussion, sometimes disagreement such as noted in the first quote at the start of this tale. But careful attention to setting things up right can make this aspect of construction straightforward and relatively carefree.

Before you get embroiled in the thankless task of payment certification, check that it is actually in your scope of work! Increasingly, clients retain quantity surveyors or other construction cost experts to perform this task. Note: it's an "all or nothing" service, meaning that you are either doing it or you are not! I have seen projects where several people were ostensibly certifying payments – a complete nightmare if there is ever a legal claim around amounts paid.

The balance of this tale will assume you are the payment certifier.

Best practices: Although most construction bids are a single number, the total cost of the project as designed, that single number is arrived at by a complex process that involves Contractors canvassing the marketplace, identifying the most advantageous combination of prices, adding them together and allowing for the Contractor's own overhead and profit.

Division	Trade	$ Bid	Work to Date	%	Last Month	%	This month	%	Remaining	%
01	General Conditions	$81,000.00	$16,200.00	20.0%	$8,100.00	10.0%	$8,100.00	10.0%	$64,800.00	80.0%
02	Excavation	$27,000.00	$25,000.00	92.6%	$13,000.00	48.1%	$12,000.00	44.4%	$2,000.00	7.4%
	Soft Landscape	$4,500.00	$0.00	0.0%		0.0%		0.0%	$4,500.00	100.0%
	Hard Landscape	$7,200.00	$0.00	0.0%		0.0%		0.0%	$7,200.00	100.0%
	Site furniture	$3,600.00	$0.00	0.0%		0.0%		0.0%	$3,600.00	100.0%
03	Concrete	$65,000.00	$8,000.00	12.3%	$0.00	0.0%	$8,000.00	12.3%	$57,000.00	87.7%
	Rebar	$42,000.00	$5,000.00	11.9%	$0.00	0.0%	$5,000.00	11.9%	$37,000.00	88.1%
04	Concrete block	$18,000.00	$1,000.00	5.6%		0.0%	$1,000.00	5.6%	$17,000.00	94.4%
	Brick	$45,000.00	$0.00	0.0%		0.0%		0.0%	$45,000.00	100.0%
05	Misc. Metals	$16,000.00	$500.00	3.1%		0.0%	$500.00	3.1%	$15,500.00	96.9%
	OWSJ	$34,000.00	$0.00	0.0%		0.0%		0.0%	$34,000.00	100.0%
	Steel deck	$27,000.00	$0.00	0.0%		0.0%		0.0%	$27,000.00	100.0%
06	Rough carpentry	$8,000.00	$800.00	10.0%		0.0%	$800.00	10.0%	$7,200.00	90.0%
	Finish carpentry	$18,000.00	$0.00	0.0%		0.0%		0.0%	$18,000.00	100.0%
07	Waterproofing	$7,000.00	$3,000.00	42.9%		0.0%	$3,000.00	42.9%	$4,000.00	57.1%
	Damproofing	$2,000.00	$2,000.00	100.0%	$1,000.00	50.0%	$1,000.00	50.0%	$0.00	0.0%
	Metal Cladding	$32,000.00	$0.00	0.0%		0.0%		0.0%	$32,000.00	100.0%
08	HM Doors	$9,000.00	$0.00	0.0%		0.0%		0.0%	$9,000.00	100.0%
	Windows	$19,000.00	$0.00	0.0%		0.0%		0.0%	$19,000.00	100.0%
	Entry Doors	$3,000.00	$0.00	0.0%		0.0%		0.0%	$3,000.00	100.0%
09	Carpet	$5,000.00	$0.00	0.0%		0.0%		0.0%	$5,000.00	100.0%
	Tile	$4,200.00	$0.00	0.0%		0.0%		0.0%	$4,200.00	100.0%
	Painting	$9,900.00	$0.00	0.0%		0.0%		0.0%	$9,900.00	100.0%
14	Hydraulic Elevator	$28,000.00	$0.00	0.0%		0.0%		0.0%	$28,000.00	100.0%
15	HVAC	$54,000.00	$0.00	0.0%		0.0%		0.0%	$54,000.00	100.0%
	Plumbing	$12,000.00	$3,000.00	25.0%	$1,000.00	8.3%	$2,000.00	16.7%	$9,000.00	75.0%
16	Electrical Rough-In	$20,000.00	$3,800.00	19.0%	$2,000.00	10.0%	$1,800.00	9.0%	$16,200.00	81.0%
	Electrical Finish	$18,000.00	$0.00	0.0%		0.0%		0.0%	$18,000.00	100.0%
	TOTAL	$619,400.00	$68,300.00	11.0%	$25,100.00	4.1%	$43,200.00	7.0%	$551,100.00	89.0%

Figure 78- Typical Contract Price Breakdown & Month 2 Claim

Once a bid is accepted, regardless who is the payment certifier, the Consultant team usually looks at the contractor's breakdown of that bid price (the schedule of values) and accepts or challenges it. That would be the "$ Bid" column above.

Occasionally a contractor will "front end load" a contract price breakdown, putting more money into the early work of the contract than the scope of work to that point merits. There is a fairly practical reason why a contractor may want to do this, which has more to do with project cash flow than any apparent dishonesty. Getting a little "ahead of the curve" creates some flexibility; being behind can create tensions with trades who are expecting payment. Sometimes agreements between contractors and their subcontractors have a "pay when paid" clause that delays subcontractor pay-ment until the contractor is paid, for obvious reasons. Even the regular time period between collecting subcontractor invoices, assembling them into an application for payment, awaiting consultant review followed by owner review before payment can easily stretch to a month and often more on larger or some institutional projects with complex owner payment approval processes.

It is important that each consultant reviews the initial contract breakdown with those thoughts in mind; be prepared to challenge the breakdown if it appears unreasonable. "Challenge" usually translates to "May I please have a breakdown of the HVAC costs; they appear excessive." Much of the time the breakdown will support the dollar amount being challenged; the consultant team will need to negotiate a change to the breakdown in the remainder of the questioned cases.

The sample above is organized by specification division, which is one standard approach – you will see many variations. After (perhaps) some adjustment to the breakdown, the results become the basis of a monthly bill from the Contractor, in which the percentage complete for each trade is tabled, together with the Contractor's own allowance for overhead and profit.

In the example above, for month two of a ten-month project duration, the "General Conditions", which are the contractor's overhead and profit, are charged at 20%, that is 10% per month. This covers items such as Superintendent salary, temporary services such as a site trailer and fencing, etc.

An experienced claim reviewer will infer from this claim that excavation is complete except for some incidentals; work has started on concrete foundations, including their damproofing; some block work has started, probably around the elevator core; and that some underground plumbing and electrical services have been installed. The point of reviewing this claim on site is to be comfortable that the proportion claimed approximately matches the actual state of the work.

After acceptance by the consulting team, the architect/payment certifier prepares a Certificate for Payment to the Owner, advising the Owner to pay $X. We will talk about the mechanics of this a bit later.

Although contracting methods such as Construction Management can become very complex for contractors and owners, the general approach has been simplified – rolled up to collective totals where the architect or another consultant is expected to certify payments. The reason for this is NOT that architects do not understand the costs of construction, although some contractors and owners might argue that. The simplification of payment certification is actually designed to protect the details of how the cost of construction is assembled, i.e., how the mutual relationship between the contractor and the trades created the winning bid.

This concealment does not arise because there is anything untoward occurring; rather, it allows contractors to preserve their business relationships with their subcontractors and suppliers, including proprietary combinations of pricing that may offer cost advantages between a general contractor and her/his subcontractor(s), resulting in the best price for the client. Longstanding "preferred pricing" between contractors and subcontractors/suppliers is known as "competitive edge" and is vigorously protected as confidential!

This is actually no different than consulting arrangements, where a subconsultant may legitimately offer a reduced fee to one architect over another, simply because the price advantaged architect is, say, more efficient hence more profitable to work with. Similarly, architects may offer a better fee for a similar project to one client over another, simply because the price advantaged client is also more efficient, or fair, or reasonable, hence more profitable to work with.

On the face of it, this may sound contrary to recommended tariffs of professional fees, where these exist. However, when you read the "fine print", such tariffs are generally guidelines based on the historic cost of services. They cannot consider the particulars of each client. What is more professional is to ensure that the nature and extent of professional services meet a certain minimum standard, for the protection of the public and clients.

Back to payment certification! The figure below is a gross simplification of the type of detail that a contractor might include in the details of an estimate – essentially, who does what to whom and for whom:

Division	Trade	Comments
01	General Conditions	Contractor Overhead & Profit - 10 month duration
02	Excavation	
	Soft Landscape	
	Hard Landscape	Patios, sidewalks, etc.
	Site furniture	Benches, bike racks
03	Concrete	Crane included
	Rebar	subcontractor to concrete supplier
04	Concrete block	same company as brick mason
	Brick	Includes base flashings & accessories
05	Misc. Metals	Includes handrails and elevator support beams
	OWSJ	
	Steel deck	erection is by OWSJ subcontractor
06	Rough carpentry	includes framing elevator opngs in block
	Finish carpentry	
07	Waterproofing	
	Dampproofing	
	Metal Cladding	Includes air and moisture barriers
08	HM Doors	
	Windows	Includes window flashings

Figure 79- Some Contractor Assumptions in Cost Breakdown

Some of this was mentioned in our tale about bidding, but perhaps needs greater explanation here because of its importance for payment certification.

Issued for construction (IFC) documents typically comprise drawings, specifications, addenda and post tender addenda (although addenda information may have been incorporated into the working drawings and specifications post award for a tidier package to build from). IFC documents do not tell the contractor how to divide and assign the work, or which trades are to do each aspect of the work – in fact, construction contracts explicitly leave the selection of subcontractors and suppliers to the contractor as well as the "means and methods" they use to get the project built. I put "means and methods" in quotations because that is the language contractors use, and therein lie the secrets of building a building, as compared with designing a building.

For example, if we think of a multi-storey project with a concrete frame, chances are the concrete subcontractor (which may, in fact, be separate sub-subcontractors for formwork, reinforcing steel (rebar) and concrete placement and finishing) will often be expected to provide a construction crane as part of their means and methods of executing the work. It will probably be included in the subcontract between the concrete subcontractor and the general contractor that the crane will be used for the movement of formwork and rebar. Depending upon the other subcontractors selected and their negotiations with the general contractor, the general contractor may also want the crane used for movement of, say, HVAC ductwork and air handling units, curtain wall or other building envelope elements, etc. The scheduling and payment for crane usage other than for concrete will involve negotiations in a series of "buy out" meetings between the general contractor and several subcontractors, none of which the architect/payment certifier should care about. As a result of these negotiations, there may be multiple extras and credits internally between the general and various subs, which the architect/payment certifier will never see (Thank God!); these will not likely affect the general contractor's proposed schedule of values, with its simple divisional price breakdown. And frankly, the architect/payment certifier does not care provided the work is well executed on schedule.

Lest contractors reading this e-book think this is all self evident, know that the chances are less than10% that any given consultant I talk with will have ever heard the term "buy out," and only a fraction of that 10% will know any of the mechanics of a buy out meeting. No kidding.

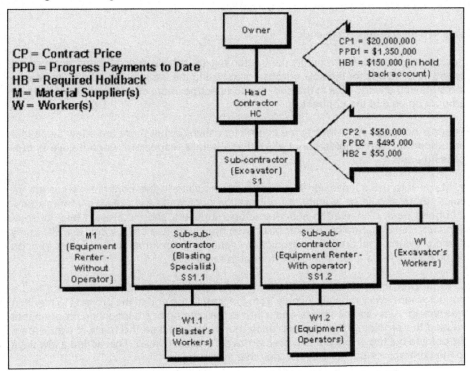

Figure 80- What could be simpler than digging a hole?

1　　　Courtesy British Columbia Law Institute, Builders Lien Act Chapt. 10, http://www.bcli. org/law-reform-resources/builders-lien-act/chapter-10

Now multiply this buy out exercise 20-50 times, which is the possible number of subcontractors on a typical project (in a large, complex project, the number of subcontractors and suppliers can easily exceed 100). The combinations are staggering and explain why good construction project managers are better paid than good project architects. The architect/payment certifier is keeping track of a few numbers within a simple breakdown; the contractor is tracking dozens.

This can get muddy quickly if the contractor starts challenging the architect/payment certifier's assessment of the work completed. Hence the first quote at the start of this tale.

But remember that the original bid for the project was probably one single dollar figure; the breakdown is for the purpose of assisting in certification. Regardless what the breakdown is, the architect/payment certifier is usually only certifying one dollar figure as payable to the contractor. The breakdown may be relied on but is not binding on the architect/payment certifier.

This latter concept is important when a contractor challenges a proposed certification dollar figure in the fashion of the quote that started this tale. Since the certifier has no real control over how that single dollar figure she/he certified is actually dispersed, the response to "I already paid 45% to the mason" only needs to be "A lot more than the mason's claim was certified. How you decide to divide up the certified amount among your subcontractors is up to you!" In other words, the contractor can decide what size pieces are given from the pie you have certified.

This is a hard argument to counter.

Back to the details. The monthly contractor invoice is broken down per the approved schedule of values for the sole purpose of assisting the various consultants reviewing the state of the work – but in the end, the architect/payment certifier recommends one amount be paid to the contractor.

There is no magic to reviewing the contractor's invoice, but there are a few steps that are important. Keep notes about all of these details, especially when delays in processing arise:

1. Make sure the submission is complete. In addition to the contractor's current invoice, there should be a statutory declaration confirming the previous month's payment has been disbursed to subcontractors, suppliers and installers. Many jurisdictions also have mandatory worker insurance premiums (we call it WorkSafe BC in my home province) that are paid periodically, resulting in confirmations or letters of good standing that need to be included. More about that later.

2. Before anything else, check the math on each contractor's invoice. I once had a project where every monthly invoice from the start to the end of the project (13 months) was wrong. I was advised there was a flaw in the contractor's accounting program that caused the problem. Nonetheless, each month we checked the math, it was wrong, we pointed out the errors and a corrected invoice was provided. This added a few days to the contractor's payment cycle, their fault not ours.

3. The second part of checking the math is to verify that the invoice only includes what has been approved as part of the construction contract. Each time there is a change to the work involving an extra or credit amount, this becomes part of the construction contract when approved by the client, not before. Regardless why a contractor might have proceeded with work that has not been approved (and there are many good reasons), if you certify payment for work that is not yet part of the construction contract you will be in trouble if, for example, the contractor leaves the project for any reason. Deleting unapproved extras is never a popular move if, for expedient reasons, that work has proceeded, but it may cause the contractor to sit with the client and iron out approvals. In any event, simply cross through unapproved items and advise the contractor you have done so.

4. Once you have verified the math and deleted any unapproved work, distribute the contractor's invoice immediately to all consultants, not just those whose design work is included in the invoice, and to the owner as information only[1] . I have had projects where a consultant with no money claimed against their design work scope to that point called me after receiving their copy of the invoice, advising "There should have been money claimed by now – I wonder if there is a problem?" This should result in a casual call from you to the contractor, with the same question. Believe it or not, I have worked on projects where a subcontractor's invoice was inadvertently missed by the contractor's accounting department, then by the project manager (remember, that's the person juggling dozens of agreements). The contractor may then submit a revised invoice including the missed amount and you and your consultant will have gained credibility as being both astute and fair.

5. Establish for the consultants a time frame for their review. Some construction contracts will specify a turnaround time for payment certification – it is safe to assume time is always of the essence. An experienced contractor will have scheduled a regular site meeting for a few days (sometimes hours) after distribution of the invoice, so that consultants have already had a chance to review the amounts and can perhaps verify amounts during the same site visit as for a regular site meeting. Consultants' review comments or approvals should be in writing (email is fine if it is being captured to the project records); if they are reducing the amount claimed, you will need their reasons as these reasons must be communicated to the contractor.

6. If you are not going to certify all of the contractor's claim, call to advise the contractor as soon as you have compiled your and the consultants' reviews and reasons to hand. It is always possible you have missed something. If not, the contractor may wish to resubmit the invoice, largely because it will have come through an accounting system that will need invoice adjustments. If the contractor proposes no invoice resubmission, it is perfectly reasonable for the architect/payment certifier to neatly annotate the invoice by hand, initial the changes and attach it to the payment certificate.

---

1   The standard CCDC2 contract in general use in Canada requires the client to be copied with claims for payment, for information and so that the client can begin to arrange for payment. Some jurisdictions and contracts may not require this, but it is certainly a best communication practice.

A payment certificate is a simple spreadsheet by any other name, so why not call it what it is? Until recently, payment certificates were word processing paper documents where numbers were plugged in by hand and all the math then separately checked:

Contract Summary

Original *Contract Price*		$ 3,750,000.00 (1)
Change Orders (numbers 1,2,4,6,7)	$ 43,562.00 (2)	
Current value of *Change Directives* included in the certified amount	$ 0 (3)	
Value of *Contract* on the last day of the payment period (1+2+3)		$ 3,793,562.00 (4)
Value Added Taxes at 5 %		$ 189,678.10 (5)
Total amount payable for the construction of the *Work* including *Value Added Taxes* (4+5)		$ 3,983,240.10 (6)

Figure 81- Word processed payment certificate contract status

The same information is now typically managed via spreadsheet in a fraction of the time and without math errors. If you think this is self-evident, know that while writing this chapter I met with a young intern architect over coffee. I mentioned this subject in passing. I was amazed to discover his firm was still doing Certificates of Payment as word-processed documents:

CONTRACT STATUS SUMMARY INFORMATION INVOICE TO DATE		
ORIGINAL CONTRACT PRICE	$3,750,000.00	
TOTAL CHANGE ORDER ADDITIONS TO INVOICE DATE	$43,562.00	CO#1,2,4,6,7
TOTAL CHANGE ORDER DEDUCTIONS TO INVOICE DATE	$0.00	
NET CO CHANGES (ADDITIONS LESS DEDUCTIONS)	$43,562.00	
REVISED CONTRACT PRICE TO INVOICE DATE	$3,793,562.00	
PLUS GST 5%	$189,678.10	
TOTAL CONTRACT PRICE TO INVOICE DATE, INCL.	$3,983,240.10	

Figure 82- Spreadsheet payment certificate details

Notice in the above spreadsheet I have specifically noted which Change Orders (1,2,4,6,7) are included in the certificate. These are the CO's that have been approved by the Owner.

Another part of the same spreadsheet details what's due this month:

TOTAL CLAIM TO DATE		$424,500.00
LESS BUILDERS LIEN HOLDBACK	10%	$42,450.00
LESS DEFICIENCY HOLDBACK		$0.00
NET TOTAL CLAIM		$382,050.00
LESS PREVIOUS CERTIFICATES		$147,256.00
AMOUNT OF THIS CLAIM		$234,794.00
PLUS GST	5%	$11,739.70
TOTAL AMOUNT DUE		**$246,533.70**
Brian Palmquist Architect AIBC		5/18/2014
(AUTHORIZED SIGNATORY)		(DATE)

Figure 83- What owed this month?

Because electronic spreadsheets can get lost, I attach them to an i-Action (i-A) for record keeping and tracking. All of the correspondence associated with a payment certificate (reviewer comments, contractor responses, etc.) is then automatically captured by QW. [1]

However you accomplish it, you should record for each certificate:

D	B	C	N/A	Item/ Information/ Question	Comments/ Conclusion/ Action
	x			Draft invoice submitted	
x				Invoice reviewed for math & correct changes	Only approved changes to the work are included.
x				Invoice returned to Builder for correction & resubmission if req'd	Establish return deadline and extend review period same amount
	x			Invoice resubmitted	
x				Invoice copied to Consultants & Client for review	Identify deadline
x		x		Consultant/Client review	Comments in writing
x				Advise Builder invoice status	Reviewed/ reviewed with revisions - detail any revisions
	x			Builder may again revise invoice	Not required but most builders will wish to align their accounting with payment certificates
x				Issue payment certificate to client	Include supporting documentation
		x		Pay contractor based on certificate	

D = Designer; B = Builder; C = Client; N/A = Not Applicable at this time; Item = Item to be considered by Designer, Builder and/or Client; Comments = by D, B, C or others. Insert the following notations as appropriate: NC = Non-Conformance; Def = Deficiency; P= Photo attached; FR = Field Review required; RFI = Request for Information required; xx = not equired of this party; 0 = information required; x = complete/signed off.

Figure 84- Payment Certification Process i-A

Then repeat each month until the project is completed, EXCEPT towards the end of the project as considered in the next tale.

---

1   This automatic aggregation of email traffic to originating i-A's is key to ISO 9001 compliance and is a feature still not widely available.

RECORD the journey		Summary Principles & Tools	
RESOLVE the issues	Be clear & quick about issues affecting certified amounts	**i-WorkPlan (i-WP)**	Include specific and custom review procedures
REVIEW the results	Check the math & the documents included with the claim	**i-Actions (i-A)**	Use i-A's to capture the process each month, a simple spreadsheet to capture the details
REMEMBER & learn to improve	Identify client-specific approval procedures	**i-KnowHow (i-KH)**	

LINKS: To access the Procedure "Review Claims for Payment" go to http://bit.ly/aagc26-1 . My work Instructions are with the "Info" link at top right. A sample "old fashioned" payment certificate is at the bottom, together with my payment certificate spreadsheet.

## 25

Paying Money
How Much Is Left?

"Is there still enough left to complete the project if the Contractor goes bust tomorrow?"
– Principal to Junior Architect

What's the point? For the first 70-80% of a project's duration, it's all about "What's the value of work completed to date?" But as each trade's work scope gets closer to the end, the architect/payment certifier needs to change mindset to "Is there enough money to complete the work?" Hence the quote at the start of this tale.

By the 70-80% mark, some trades will have accumulated a group of issues, nonconformances and deficiencies that need to be attended to before completion. Is there enough money to complete them together with the scope of work not started yet? Sometimes not. For example, consider the claim for payment tabled in the previous tale. Notice there is only $2000.00 remaining in the excavator's contract. It's reasonable to verify BEFORE paying this claim that $2000.00 is sufficient for the excavator to complete, including any deficiencies.

What are the principles & best practises? The architect/payment certifier will need to reverse mindset at this point in the project. This may occur over a period of months as various subcontractors approach the stage in their efforts where "cost to complete work and deficiencies" overshadows "percentage complete to date." Even with advance warning, the contractor and subcontractors will be shocked when you start to reduce certification based on cost to complete.

To respond to arguments around cost to complete, you will need to have been tracking issues, nonconformances and deficiencies continuously throughout the project. We will talk further about how to do that in the Tale about "Non-Conformances & Deficiencies."

RECORD the journey	Identify when each trade approaches 70%-80% completion	Summary Principles & Tools	
RESOLVE the issues	Track unresolved issues involving costs	i-WorkPlan (i-WP)	Specific procedure as part of payment certification
REVIEW the results	Review the total value of deficiencies and unresolved issues against amounts claimed	i-Actions (i-A)	
REMEMBER & learn to improve		i-KnowHow (i-KH)	

## 26

Paying Or Not?
Criminal Behaviour - Statutory Declarations

"I wondered where that seal got to." – Notary to Architect

What's the point? So you've gone to the trouble of carefully calculating how much work the contractor should be paid for each month – how do you know all of the sub-contractors, suppliers and installers are being paid? The fact is, you don't, but construction contract law in most places has invented the Statutory Declaration to give you some peace of mind.

What are the principles? Each month after the first month of construction, and at the end before holdback release, the contractor is required by most contracts to prepare a "stat dec" (as it is inevitably shortened to):

Statutory Declaration

Standard Construction Document

of Progress Payment Distribution by Contractor   CCDC 9A - 2001

To be made by the Contractor **prior to payment** when required as a condition for either:

☐ second and subsequent progress payments; or

☐ release of holdback.

Identification of Contract

The last application for progress payment for which the Declarant has received payment is No. _____

dated the _____ day of _____ ,

in the year _____ .

Figure 85 - Sample Stats Doc (top)[1]

The key element of the Stat Dec is the contractor's declaration, under oath before a notary, lawyer, barrister or attorney (depending upon jurisdiction), that accounts covered by the previous month's payment have been paid in full to the subcontractors:

Declaration

I solemnly declare that, as of the date of this declaration, I am an authorized signing officer, partner or sole proprietor of the Contractor named in the Contract identified above, and as such have authority to bind the Contractor, and have personal knowledge of the fact that all accounts for labour, subcontracts, products, services, and construction machinery and equipment which have been incurred directly by the Contractor in the performance of the work as required by the Contract, and for which the Owner might in any way be held responsible, have been paid in full as required by the Contract up to and including the latest progress payment received, as identified above, except for:

1) holdback monies properly retained,
2) payments deferred by agreement, or
3) amounts withheld by reason of legitimate dispute which have been identified to the party or parties, from whom payment has been withheld.

I make this solemn declaration conscientiously believing it to be true, and knowing that it is of the same force and effect as if made under oath.

Declared before me in _____ this _____ day of _____ ,

_____ *City/Town and Province*

in the year _____ .

_____   _____
Signature of Declarant   *(A Commissioner for Oaths, Notary Public, Justice of the Peace, etc.)*

Figure 86 - Typical Stat Dec Declaration wording[2]

---

1   CCDC 9A – 2001, from ccdc.org. Note, the use of the full form requires purchase of a seal from CCDC. This is a portion of the form and is non-editable.
2   CCDC 9A – 2001, Ibid.

To underline the seriousness of a false declaration:

> **The making of a false or fraudulent declaration is a contravention of the Criminal Code of Canada, and could carry, upon conviction, penalties including fines or imprisonment.**

Figure 87 - or else!

What are the best practices? The quote at the top of this sidebar arose on a fellow architect's project. He noticed that, although there were many notaries in the town where his project was being built, the notary seal was from a town an hour's drive away. He called the notary to inquire, was advised the notary's seal had been stolen and since replaced. My colleague's next call was to the Royal Canadian Mounted Police. Fraud charges against the contractor ensued. An extreme example, but it underlies the reliance on the statutory declaration by everyone in construction.

As an architect/payment certifier, you are expected to receive and review an original statutory declaration with original signature and seal on it, simply verifying it is for the correct month and the correct project. You may scan it for your own records, but it should go as an original with the Payment Certificate to the Owner, who should have it reviewed by legal counsel.

Another tale in this e-book will describe a project of mine where the contractor declared bankruptcy and skipped town. Fortunately, I had been scrupulous about statutory declarations, so was not liable to any of the (many) unpaid subcontractors and suppliers. Clearly the project's stat dec's were fraudulent, but determining that the apparently properly prepared and sworn stat dec's were lies was not my responsibility. They were properly prepared and sworn, so there was no claim of negligence against me. Phew!

RECORD the journey	A stat dec for each payment request after the first	**Summary Principles & Tools**	
RESOLVE the issues		**i-WorkPlan (i-WP)**	Part of the specific payment certification procedure
REVIEW the results	Look for original documents, signatures and seals	**i-Actions (i-A)**	Transmittal to client
REMEMBER & learn to improve		**i-KnowHow (i-KH)**	

## 27

Investigating the Root Cause of Recurrent Problems

"I hope you're right that that will never happen again, but I need to know why the beams cracked before I can be comfortable it won't occur again."
– Structural engineer to contractor

What's the point?Two large steel beams had cracked during the cambering process (that's where the beams are bent up in the middle so they will settle level when dead and live loads are superimposed). They were set aside for recycling and new ones were ordered. More details below.

What are the principles? During the course of construction patterns of problem may emerge. For example, wall insulation may be regularly in a heap on the floor when insulation reviews are scheduled. Or each and every one of the first five pre-drywall reviews may fail.

These are simple occurrences that are usually easily resolved. They often come down to the builder getting ahead of him/herself and sequencing work incorrectly, so installed work has to be removed or damaged to accommodate materials scheduled for later installations. The designer, in reviewing work in progress, has every right, will in fact benefit from advising the builder when there appear to be recurring issues. At best it is a waste of the designer's time and effort in the field; at worst it may result in downstream litigation against the designer and builder both if an uncorrected pattern persists and is not caught and corrected before project completion.

Sometimes the pace of construction is so frenetic and/or the construction site is so expansive (think shopping mall) that the superintendent may not notice a pattern that is evident to the professional focused on review in the field. In this case the superintendent will usually thank the consultant for pointing out the problem, and resolve it quickly. In rarer circumstances, a recurrent problem may arise that has no simple explanation. In these cases root cause analysis (RCA) may resolve the issue.

What are the best practices? The questions in the RCA i-Action below are very straightforward and derived by me from unnecessarily more complex approaches. These questions are often not asked, or asked out of sequence. The fast pace of construction impels builder field staff to want to just "get on with it" – they are not trying to hide a problem, they simply do not want to delay construction while the cause is properly scoped out and managed.

If there is resistance to determining the root cause of a problem that might be serious/ recurrent, the designer must defend the public interest by insisting on a pause while the root cause is determined and any measures taken that will prevent a recurrence. Just insisting that the questions below be considered and answered in a rational fashion will often allow for reason to prevail.

The root cause analysis i-A below is a succinct description of the classic RCA process:

To reply or add further information, either login to Quality-Works.net, or just hit the 'Reply All' button, then add notes in the appropriate table cell and 'Send.'

ROOT CAUSE ANALYSIS (RCA) REPORT - [SUBJECT]

This is a summary of the investigation and recommendations associated with an event requiring root cause analysis. Refer to the bottom of this form for instructions on its use.

Ref #	Item
1	Definition of Problem:
2	Evidence/ data gathering:
3	Apparent cause(s) of defined problem:
4	Which cause(s) if removed or changed will prevent recurrence:
5	Effective solutions that prevent recurrence, are within our control, meet our goals and objectives and do not cause other problems:
6	Recommendations for implementation:
	Report prepared by (names/ positions/ contact info):
7	Review of implemented recommendations (by report writers and others):

Instructions for use:

- It is acceptable to prepare, issue, circulate and fill in this form by hand, although it will be easier electronically.
- The person issuing this report is the first person indicated as 'Referred By.'
- The Contractor staff member responsible for reviewing or implementing this report is the person indicated as 'Now Responsible.'
- The Contractor staff member responsible should cc this report to those who need to know the root causes and implement the recommendations.
- Indicate the type of installation reviewed using the 'Element' drop down list at the top.

Figure 88 - Root Cause Analysis Report i-A

The party responsible for the event(s) being investigated usually pays for Root cause analysis. Regardless who is paying, the consultants need to be comfortable that the process is independent. Chances are the responsible party will be the contractor but the RCA process itself may involve consultants or the owner if they witnessed the failure event(s).

A good example of root cause analysis involving consultants would be the unexpected failure of a building component test. Wearing my building envelope consultant hat, I once witnessed the unexpected failure of a huge section of curtain wall at the manufacturer's test facility. We were testing its reaction to hurricane force winds when there was a huge bang and all the pressure gauges cycled to zero.

Nothing appeared broken from a distance, but from close up it appeared several bolts fastening the curtain wall to the test chamber had failed, causing the entire test section (two storeys high and four windows wide) to move within the test frame. Fortunately, the construction of the test frame itself prevented the curtain wall from moving more than about 4" (10cm); there were workers and consultants behind the test panel in the direction it moved (not me, I was in a control room witnessing pressure gauges).

Investigation revealed that due to the absence of the specified gauge of bolt to temporarily fasten the curtain wall to the test frame, a smaller gauge had been used but not noticed, which was not strong enough for high wind forces. So there was nothing wrong with the curtain wall itself, as retesting the next day showed – it was just the wrong bolts being used for the temporary testing apparatus.

As demonstrated above, data gathering and observation in situ are probably the most important aspects of the root cause analysis process, and the sooner the better. Memories dim and are confused by multiple reports, third party comments, etc. It is essential to get to those who witnessed the events being investigated as soon as possible – their information and recollections will often contradict the guesses that will have already arisen around events.

Considering the cracked beams mentioned at the start of this Tale:

Definition of Problem – Construction sites are rumor mills. Many times the actual problem is not the problem that you or others have heard. Third hand knowledge is reported as fact. Keep editorializing out, simply define the problem – "The beam cracked," "water came in through the window sill", etc. Argue as long as it takes to agree on a cogent statement of the problem. Bring in experts as necessary to assist with problem definition.

Evidence/data gathering – The damaged or defective beams were set aside in a protected area and marked not to be used elsewhere - all good. It is important the builder mark "Do not Use", otherwise the material may appear perfect for another installation. The protection is also important in case the damage is affected by weather.

In the case of the beams, the contractor hired a materials science expert who first interviewed the persons who were at hand when the beams cracked. It does no good to have a report from anyone but those who were on hand when the problem occurred. Again, you are not seeking opinion, simply statements of fact and recollection – "It broke as soon as we put weight on it", "It looked crooked out of the package", etc.

A knowledgeable expert will know to ask questions like "How cold was it when the accident occurred?", "How long had it been outside before it was brought in to work on?", etc.

"The facts, just the facts, Madam!"[1]

Apparent cause(s) of defined problem – Here is where the true analysis begins. Depending upon the nature of the problem, it may be necessary to engage experts to determine how widespread a problem might be, as it was in the case of the cracked beams. The problem could affect only one or two items or, in the worst case, an entire shipment. It could involve one window or every seal on every window. It will be impossible to decide how to check the extent of the problem until a cause has been established.

Our expert followed the steps noted above and in the i-A. He concluded the cracking occurred because holes in the two beams in question had been punched when they should have been drilled, and before rather than after cambering. There was nothing in the construction documents covering this particular sequencing of treatment. We learned that there seldom is. The expert reviewed mill tests, examined the steel with an electron microscope, examined weather records, the equipment used to punch and camber, etc. He concluded it was an unfortunate combination of circumstances that would have been avoided had the beams been drilled instead of punched and after rather than before cambering. But the order and nature of the work was altered and the work proceeded without further incident.

---

1   after Sergeant Friday

The root cause of the failure was very different than first guessed. Identifying the cause resulted in a simple change to the procedures for preparing beams – no additional beams cracked and the project team learned important lessons about beam preparation.

Such is root cause analysis at its best. It may never be required on your project, but if it is, take it seriously.

RECORD the journey	Capture any unusual occurrence	Summary Principles & Tools	
RESOLVE the issues	Use a rigorous RCA process	i-WorkPlan (i-WP)	Specific RCA procedure
REVIEW the results	Is there a change to procedure going forward?	i-Actions (i-A)	Root Cause Analysis Report
REMEMBER & learn to improve	Modify documents of i-WP around root causes	i-KnowHow (i-KH)	Capture i-KH's for each event, for ease of future search

LINKS: To access the procedure around root cause analysis, go to http://bit.ly/aagc27-1

28

Change happens - Manage the Risks

"This excavation is a nightmare! Thank God the owner agreed to a large foundation contingency. We're going to have a forest of piles!"
– Construction Manager to Architect.

What's the point? Change happens in design and construction. Clients change their minds or have their minds changed by real world circumstances. Regulations change, sometimes retroactively. Unexpected things are found when a hole is dug, or when an existing wall is torn down or torn open. But you can manage change such that it is neither unpleasant nor unrewarding.

In many construction contracts except some lump sums, the builder will specifically identify the cost of risk as one or more contingencies, carry that amount in the schedule of values used as the basis of monthly invoices, and charge against it as circumstances arise. Good bid documents will identify unknowns that the designers are aware of, and specifically recommend that allowances be declared and carried in the bid.

Some designer contracts include contingency amounts, but less frequently and even more less frequently in these fiscally tight times. But the designer still has some means of minimizing risk associated with changing circumstances.

What are the principles? The key to minimizing risk is to take the builder approach. In a lump sum environment, the builder has no idea about risk until the estimating process is started. But each estimator, including estimators for subcontractors and suppliers, will pore over the bid documents, geotechnical reports, etc. and identify the potential for issues to arise that are not covered by generic change to the work items in bid documents.

What do I mean by "generic change to the work"? As an example, because geotechnical reports are based upon the geoscientist's experience of the general area together with spot holes drilled and core samples taken, it is wise to allow for the possibility of "your model may differ" conditions. These may be covered by a bid requirement for unit prices for certain types of construction (so many $ extra for deep piles compared to specified spread footings), or simply by negotiation based upon marketplace quotes for extra work.

To illustrate how fickle geotechnical reports can be, I was project architect and designer for a private care home in Victoria, British Columbia some years ago. Because of its route on the great glacier advance and retreat routes, soils in Victoria can vary hugely just feet apart. The geotechnical report for our project was well done, and each test hole identified good bearing only a few feet below grade, meaning we could anticipate the relatively inexpensive spread footing approach to foundations.

As we began to excavate we discovered that, sure enough, the soils under every test hole were firm starting a few feet below grade. Unfortunately the site was like a miniature mountain range, with gullies of soft soils a few yards from our test holes. In the end we had a foundation consisting of about equal parts spread footings and deep piles. For the less experienced among you, deep piles require deep pockets, i.e., are expensive.

Fortunately the builder had negotiated a contract with the owner and had argued successfully for a substantial foundation contingency. The owner was unhappy about spending money on unseen supports below grade, but the project remained financially viable because money for deep foundations had been included in the construction budget.

What are the best practices? The first key to managing change is to identify known risks in your project. Most experienced builders will have their estimators identify risks during the estimating/bid stage of a project. Potential costs will be set against the unknowns, called contingencies. If builders do not identify risk in a methodical fashion, their insurers will probably make them – or they will soon go out of business. The i-A table below is recommended by a major construction insurer:

Project Risks

This form is used to transmit & evaluate project risks. To complete this form or reply to this email, select "Reply All", update information & select "Send". "B" = Builder. "N/A" = not applicable on this occasion; "D" = Designer. if you are using an iPhone or iPad you can select a text box and dictate comments.

D	B	N/A	Item	Solution	Remarks

Rev 001 - 140923

Figure 89 - Project Risks i-A

Another way to track project Risks is to capture each with its own i-A and keep it open until the issue is resolved:

Action								
Now Responsible:			Referred By:					
Item:	Drywall quality risk - sulphur		Action Number:	A00160 R0		Priority:	11 Reference	
Type:	Issue		Subtype:	Architectural		Element:		
Status:	Open		Due:	2014/05/09		Opened:	2014/05/08	Closed:
Identifier:			Originator:					

D	B	N/A	Item	Solution	Remarks
			There have been increasing reports of drywall imported from Asia having sulphur as one of the core ingredients instead of gypsum - the product is very corrosive when wet, may also emit noxious odours		
x				Specify a chain of custody for drywall supply & require all supplies come from a North American company	This issue is real and arose on one of our projects. Because it was identified at the design stage, specifications clearly prohibited the suspect product - there were no issues during construction
	x			Provide chain of custody & verification product manufactured locally	

Figure 90 - Project Issue as Individual i-A

The Project Risks list will ebb and flow during the course of designing, estimating and building a project. Once construction starts, the construction contract will include a more formal change management process based on identifying the nature of a change, determining if it has cost or schedule impacts, quantifying those impacts, obtaining agreement to impacts from the client, contractor and consultants, then proceeding with the change. Sounds simple, so why is it so often so problematic?

Going forward following that Victoria deep foundations project, when preparing an i-WP for a Victoria project our "Victoria" tag (Figure 7) includes a procedure that recommends a deep foundation unit price or contingency be included in the bid documents.

RECORD the journey	Log each project risk	**Summary Principles & Tools**	
RESOLVE the issues	Or use i-A's to capture and resolve each project risk	**i-WorkPlan (i-WP)**	Specific procedures
REVIEW the results	Keep reviewing the risks during the course of design & construction	**i-Actions (i-A)**	Risk Review log or separate i-A for each risk
REMEMBER & learn to improve	Refine i-WP's	**i-KnowHow (i-KH)**	

LINKS: For a copy of my risk review procedures and form, go to http://bit.ly/aagc28-2

29

Change happens – Manage the Details

Setting aside the client's reluctance to pay any more than the original contract amount, the consultant's desire to build exactly what was designed and the contractor's desire to be fairly compensated for every change, change management has historically been straightforward but time consuming.

Change in construction is generally managed via four means: site instructions; contemplated change orders; change orders; and change directives.

A Site Instruction (SI) is a clarification of the contract documents that the issuing consultant does not believe affects either cost or schedule. Think of the contractor asking for a window jamb detail and the architect responding with a minor variation to an already-drawn detail.

> SITE INSTRUCTIONS are issued only for the purpose of recording any clarification or interpretation of the contract documents or giving direction on problems resulting from field conditions. These instructions are subject to the provisions of the contract documents and unless stated herein and specifically co-authorized by the Client, will not affect the contract as to value or duration. Should the Contractor wish a change to the contract price or schedule, he/she shall submit a claim to the Architect within ten (10) days from the date of this Site Instruction.

Figure 91 - Site Instruction definition

When a site instruction is issued, the contractor usually has about ten days by contract to advise if the SI does, in fact, merit a change in contract cost or time. If that is the indication, there may be some discussion or debate. Let's assume the architect is convinced that adjustment of cost or schedule is merited. Perhaps in the window jamb detail above, the architect realizes the identified condition is substantially different than drawn details and requires additional material for a proper installation. The architect then issues a contemplated change order (CCO) or contemplated change notice (CCN) – same thing.

> CONTEMPLATED CHANGE: This is a notice to the Contractor describing the proposed change in the Work. The Contractor shall present, in a form acceptable to the Architect, a method of adjustment or an amount of adjustment of the Contract Price, if any, and the adjustment in the Contract Time, if any, for the proposed change in the Work.

Figure 92 - Contemplated Change Definition

The CCO is just that – "contemplated." Often CCO's arise when the client decides to change the program or other details. I once worked on a project where the owner decided during construction to price out adding an entire floor to the building! We designed and documented a floor over one "all hands on deck" weekend, the contractor priced it and the owner approved it. The design fee was 6%, or $180,000 for the change. Not bad for a weekend's work!

The contractor will respond to a CCO by soliciting pricing from all affected subcontractors and suppliers – this can be a significant number of players. The contractor will assemble the pricing for the consultants' and client's review.

Incidentally, clients and consultants have little understanding of what's involved in pricing a change. There can easily be ten or more trades affected by a change, any or all of who will have other pressures and deadlines. The contractor will need to solicit input from all of them, then assemble it into a single price. For this reason, many contractors resist pricing "exercises" when they appear designed to allow the client or consultant to "try on" different ideas during construction.

Pricing proposed for a CCO may be subject to clarification, breakdown and debate. When further agreement has been reached about adjustment to time and schedule, the architect will issue a Change Order (CO) or Change Notice (CN) (same thing) to change the contract value and (often) change the contract time.

> **CHANGE ORDER**: When the Owner and the Contractor agree to the adjustments in the Contract Price and Contract Time or to the method of adjustment to be used to determine the adjustments, such agreement shall be recorded in a Change Order signed by the Owner and Contractor and effective immediately. the value of the work performed as a result of a Change Order shall be included in applications for progress payment. NOTE: All costs associated with this Change Order, including Contract Time Change, if any, shall be included in this cost adjustment.

Figure 93 - Change Order Definition

Finally, on some occasions the contractor needs to proceed with a change to the work faster than the above process allows. In this case, the Owner must agree to the issuance of a Change Directive, which authorizes work to proceed in advance of determining its cost.

> **CHANGE DIRECTIVE:** If the Owner requires the Contractor to proceed with a change to the work prior to the Owner and the Contractor agreeing upon the adjustment in Contract Price and Contract Time, the Owner, through the Architect, shall issue a Change Directive. Upon receipt of the Change Directive, the Contractor shall proceed promptly with the Change in the Work. Adjustments to the Contract Price and Contract Time shall be determined in accordance with Article GC 6.3 of the CCDC 2 Construction Contract.

Figure 94 - Change Directive Definition

Early in my career I struggled with keeping track of these four kinds of change. Now, did SI#6 become CCO#8 or CCO#10? Did CCO#10 become CO#12 or CO#13? How much do we have in outstanding CO's anyway?

Noticing the numbers were not important except as placeholders, and that the time/ budget aspects used the same math regardless, I decided "a change is a change is a change" and developed a simple spreadsheet called "Change in the Work."

Change in the Work logo here

DATE:	date
CHANGE IDENTIFICATION #	1
SITE INSTRUCTION #:	
CONTEMPLATED CHANGE #:	
CHANGE ORDER #:	1
CHANGE DIRECTIVE #:	
PROJECT NO.:	9999
PROJECT:	PROJECT NAME PROJECT ADDRESS
OWNER:	OWNER NAME OWNER ADDRESS
CONTRACTOR:	CONTRACTOR NAME CONTRACTOR ADDRESS

Distribution Instructions:
Site Instructions: can be faxed to Contractor and Owner, signatures not required (except Architect's).
Contemplated Changes: can be faxed to Contractor and Owner, signatures not required (except Architect's).
Change Order: original required. Three originals must be signed by Architect, Contractor and Owner. The Architect will forward to the Contractor first, then the Contractor forwards to the Owner. Owner must return the signed copies back to the Architect. Architect distributes one copy each to the Contractor and Owner, and one for the file.
Change Directives: original required. One signed Change Directive by the Architect must be sent to the Contractor.

Figure 95 - Change in the Work Spreadsheet Top

The key concept here is that when a change is identified, it is sequentially numbered. The illustration above is #1. It stays as #1 regardless which of the four change types it is. If it is issued as SI #1 and finishes there, then #1 is not used again on the project. This sample is Change Order #1. The next change is #2, regardless whether it is a SI, CCO, CO or CD. SI#3 may become CCO#3 and CO#3 but it will never become CO#4.

The middle portion of my "Change to the Work" provides the legal description of what this is, as well as space for coordinates such as Location in building, Description of Change, etc.:

SITE INSTRUCTIONS are issued only for the purpose of recording any clarification or interpretation of the contract documents or giving direction on problems resulting from field conditions. These instructions are subject to the provisions of the contract documents and unless stated herein and specifically co-authorized by the Client, will not affect the contract as to value or duration. Should the Contractor wish a change to the contract price or schedule, he/she shall submit a claim to the Architect within ten (10) days from the date of this Site Instruction.

CONTEMPLATED CHANGE: This is a notice to the Contractor describing the proposed change in the Work. The Contractor shall present, in a form acceptable to the Architect, a method of adjustment or an amount of adjustment of the Contract Price, if any, and the adjustment in the Contract Time, if any, for the proposed change in the Work.

CHANGE ORDER: When the Owner and the Contractor agree to the adjustments in the Contract Price and Contract Time or to the method of adjustment to be used to determine the adjustments, such agreement shall be recorded in a Change Order signed by the Owner and Contractor and effective immediately. the value of the work performed as a result of a Change Order shall be included in applications for progress payment. NOTE: All costs associated with this Change Order, including Contract Time Change, if any, shall be included in this cost adjustment.

CHANGE DIRECTIVE: If the Owner requires the Contractor to proceed with a change to the work prior to the Owner and the Contractor agreeing upon the adjustment in Contract Price and Contract Time, the Owner, through the Architect, shall issue a Change Directive. Upon receipt of the Change Directive, the Contractor shall proceed promptly with the Change in the Work. Adjustments to the Contract Price and Contract Time shall be determined in accordance with Article GC 6.3 of the CCDC 2 Construction Contract.

| REFERENCE: |
| LOCATION IN BUILDING: |
| DESCRIPTION OF CHANGE: |
| COMMENT: |

Figure 96 - Change in the Work Spreadsheet Middle

Here I have highlighted that this is a SITE INSTRUCTION. Naysayers have suggested having the four definitions bunched together will confuse contractors, but this has never occurred in my usage, perhaps because the top part of the form clearly identifies what type of change is being considered.

The subjects at the bottom, REFERENCE, LOCATION IN BUILDING, DESCRIPTION OF CHANGE and COMMENT, often never change. The latter two sometimes morph as the change evolves, never the first two – so why transcribe them again and again? Where a change originates from a subconsultant, I require that individual to add a comment explaining why the change is arising.

The actual spreadsheet math only occurs at the bottom of the spreadsheet:

ORIGINAL CONTRACT AMOUNT:	$3,750,000.00	
EXTRAS APPROVED TO DATE:	$43,562.00	(CO#1,2,4,6,7)
CREDITS APPROVED TO DATE:	$0.00	
PROPOSED EXTRA - THIS CHANGE:	**$1,867.00**	(Attach detailed Breakdown)
PROPOSED CREDIT - THIS CHANGE:	$0.00	(Attach detailed Breakdown)
OTHER ADJUSTMENTS:	$0.00	
REVISED CONTRACT AMOUNT:	$3,795,429.00	
OTHER EXTRA'S AWAITING APPROVAL	$2,568.00	(CO#3,5,8)
OTHER CREDITS AWAITING APPROVAL	$0.00	(CO#)
CONTRACT TIME:	x	
CHANGE TO TIME:	none	
DAYS ADDED TO CONTRACT TIME:	none	no additional costs
DAYS DEDUCTED FROM CONTRACT TIME:	none	no additional costs
ISSUED BY:	Architect	
ACCEPTED BY:	Contractor	
ACCEPTED BY:	Owner's Representative	

Figure 97 - Change in the Work Spreadsheet Bottom

There are a few things about the math worth noting:

EXTRA'S AND CREDITS APPROVED TO DATE lists only those items that have already been approved by the client – additional items in the pipeline but not yet approved are excluded as they are not part of the contract yet. This information is always maintained current, so even if the change is an SI, the client, consultant and contractor have a reminder of contract status.

PROPOSED EXTRA'S AND CREDITS for this change are straightforward, get filled in when the contractor's proposals have been vetted and recommended by the consultants. Incidentally, changes involving specialist consultants such as structure, mechanical, etc., should always be reviewed by them.

OTHER ADJUSTMENTS includes any items that have resulted in a contract change, but were not handled via a change in the work; perhaps a post tender addendum item, for example. Note that a post tender addendum is only executed between the client and the successful bidder – it is not a tendering or negotiating instrument.

The REVISED CONTRACT AMOUNT is the simple sum of the above six spreadsheet cells.

OTHER EXTRA'S AND CREDITS AWAITING APPROVAL is here for information. Clients sometimes forget what other financial commitments are awaiting their review and approval; some clients insist on knowing what their maximum financial exposure is. These numbers allow them to establish that exposure.

The TIME components at the bottom provide a running total of whether and how much the contract time has changed. This is important to the client if there is a fixed move in date (think first day of school). I added the "no additional costs" note as the default (can be changed) after an experience where a contractor kept agreeing to changes without charging extra so long as the schedule was extended. Then at the end he claimed for several weeks of additional general conditions (superintendent salary, site trailer, insurance, etc.) even though he had not actually spent all of the time he had banked. We settled the matter amicably but I wanted to avoid future "misunderstandings."

RECORD the journey	Log each change to the work	**Summary Principles & Tools**	
RESOLVE the issues	Use i-A's to capture discussion about changes	**i-WorkPlan (i-WP)**	Specific procedures
REVIEW the results	Carry total financial commitments in each change	**i-Actions (i-A)**	Change to the work i-A
REMEMBER & learn to improve		**i-KnowHow (i-KH)**	

LINKS: To look at a Change to the work i-A with attached spreadsheet, go to http://bit.ly/aagc29-1. Click on "Uploads (1)" at the bottom of the screen to open the change spreadsheet.

## 30

Changing Design during Construction
Who Cares?

"It's not a change to the development permit. It's just a color and material change."
– Client to Architect

"The city architect has reviewed and approved the proposed changes."
– Client to Architect

What's the point? It is easy to forget in our focus on the time and expense surrounding changes to a project that we may be changing the project's design as approved by the local authority. Variations on the above quote have arisen on several of my projects, with two basic non-monetary impacts.

What are the principles? First, there is the technical aspect of a design change. Is the proposed alternative cladding material as robust as what it replaces? If the answer is yes, then have we modified the various interface details around the new material? Our insurers and lawyers will be unsympathetic if we saved someone a few dollars by not re-detailing at the expense of a compromised building envelope.

Equally important is the regulatory aspect of design. We will probably need a building permit amendment for a new material. We will almost definitely need a planning permit amendment for either color or material changes. These can take weeks or months and in the end may not be accepted.

On several occasions when acting as a code or building envelope consultant I have quietly reminded the architect of these realities and the likelihood she/he will be in the regulatory agency's and the profession's bad books if design change is not conducted in the fashion prescribed by the authority. Perhaps because they are more attuned to seeing change as a monetary or marketing element, clients and contractors often tend to downplay the professional risks associated with premature design change. Consultants cannot. See below for details of my own nasty experience in this regard.

What are the best practices? As soon as a proposed design change of any kind is tabled, you should begin to track it, using an i-A if that is your approach. It may be abandoned along the way – if you are a consultant, your contract should identify a change to the design as being an additional paid service, and you may invoice the client for the change at the point of abandonment, using the i-A and associated correspondence and design work as backup.

If the change proceeds after regulatory requirements have been identified, resolved and recorded, then the i-A may morph into a change to the work in addition to changes to consultant documents. It is important that you record the evolution and resolution of such changes. Regardless of disposition, recording the history and resolution will provide good backup for your eventual charges.

I once made the mistake of believing the client who uttered the second quote at the start of this tale. Actually, he was in a major dispute with the city in relation to another project I knew nothing about. The city retaliated by visiting the construction site of my project after hours and videotaping modest changes I had been assured were approved. A complaint to my professional association against me was filed by the city. I was exonerated, but only after seven years of litigation, damage to my reputation and more than $100,000 of unreimbursed legal expenses (in court, even when you win, you lose).

RECORD the journey	Capture each potential design change from the moment it arises	Summary Principles & Tools	
RESOLVE the issues	Obtain necessary regulatory approvals	i-WorkPlan (i-WP)	
REVIEW the results	Identify the cost of changing against the "savings"	i-Actions (i-A)	i-A for each proposed change
REMEMBER & learn to improve	Capture specific amendment procedures for each jurisdiction	i-KnowHow (i-KH)	

31

Changing Details
"The Detail's Just the Same"

"I'm surprised the 4-2x6's have opened up. A 6x6 would have just cracked anyway." – Contractor to Architect

Figure 98 - The Changed Detail did not Work

What's the point? I was the building envelope consultant on the project. The design drawings showed a crowned 6x6 (140x140mm) (i.e., the top surface of the 6x6 had a slight slope down from the center) visually supporting the small roof over the entry door. I had commented during design review that the portion of the 6x6 that was protruding past the barge board would be subject to extra weathering. Imagine my surprise when I discovered that during construction the 6x6 had been changed to 4-2x6's (38x140mm), which had immediately opened up to invite the water in. What had started as a possible aesthetic concern now became a possible point of building envelope failure. The unfortunate wrapped metal flashing protects the wood at the expense of any charm the entry might have had.

What are the principles? A design change portrayed as "no change" by the builder is nonetheless a change. It requires design review to ascertain it has not compromised design aesthetics, code requirements or building envelope practicalities. In this case, a detail that I had identified as marginal (an end-exposed 6x6 that would be more likely to pick up water through any cracking or splitting) became unacceptable when changed to multiple plies of 2x6.

What are the best practices? When a design change is proposed, even if it appears and is presented as minor, it should be considered as a value engineering exercise such as outlined in Tale #6. In some instances, identifying the true soft costs (additional fees to the architect and building envelope consultant, in this case) and hard costs (an additional wrapped flashing in this case) may cause a "cost savings" design change to be abandoned. If not, it will first be subject to professional review, and the professionals will be reimbursed.

RECORD the journey	A separate i-A for each proposed change	**Summary Principles & Tools**	
RESOLVE the issues	Each affected consultant to review	**i-WorkPlan (i-WP)**	
REVIEW the results	Is there a cost/schedule savings in the end, or quality improvement?	**i-Actions (i-A)**	Value engineering template
REMEMBER & learn to improve		**i-KnowHow (i-KH)**	i-KH's for new knowledge

LINKS: To review a Value Engineering Report template, go to http://bit.ly/aagc31-1

## 32

Change that is NOT?

"I never issue SI's, CCO's, CO's or CD's – just Architect Instructions. They're much easier for me to understand and they seem to work out fine." –Architect to Consultant

There are some forms of contractual arrangement that do not necessarily contemplate the four types of change described in previous tales. A consistent example is when the consultant team is working for a developer who is also a builder. Since the consultants are likely not reviewing the proposed costs of changes (or any other financial matters), it may seem acceptable to the developer/builder to blend all change types together into what I will call Supplemental Instructions. Think of the architect as having become the Developer/Builder's "artistic director".

Depending upon the circumstance and the day of the week, a given Supplemental Instruction may in fact be a site instruction, contemplated change, field review, non-conformance or deficiency, even a submittal review. No wonder there is such confusion on some construction sites.

There is nothing inherently wrong with this approach in this untypical environment, but some words of caution are required. Whereas the conventional contract's clarity about which changes have been approved or not gives the consultants an accurate picture of what is being built or may be built, a Supplemental Instruction may or may not be implemented. Some builders believe that since they are more intimately related to their client/developer, they may choose after discussion with their colleague "client" (being as they may be two divisions of the same company) either to a). construct the detail as drawn; b). accept the instruction as noted; c). ignore the instruction entirely; or d). amend the instruction as they feel appropriate.

Consultants engaged in these circumstances should treat them as rigorously as they would manage conventional contract change: log each instruction and verify it has been implemented or not, especially if it arises from a public safety requirement, including from planning and building regulations.

The relaxed practices described above can create a relaxation in professional discipline within the design studio and in the field. Thereafter, re-establishing the necessary rigor to take on more stringent projects with more conventional contract agreements can be difficult, especially for junior staff who may presume that what they have been doing on developer/builder projects is "the right way" for all projects. I have personally witnessed this confusion by juniors in the field, in the office and in oral examinations to become an architect.

There are some architects who only work for developer/builders. That is a business decision on their part. It is not a prudent professional decision for intern architects.

RECORD the journey	Identify the nature of each "instruction."	**Summary Principles & Tools**	
RESOLVE the issues	Track resolution as you would for a conventional change	**i-WorkPlan (i-WP)**	Specific procedures for developer/ builder projects
REVIEW the results	Ensure instructions have been implemented	**i-Actions (i-A)**	i-A templates as for conventional contract changes
REMEMBER & learn to improve	Remember the nature of changes when working on other projects	**i-KnowHow (i-KH)**	

## 33

Evaluating Alternatives & Substitutions
Dividing the Labor

"Just look it up on their website, it's really well done. I'm sure you'll find the information you need there. Besides, they're just the same" – Contractor to Architect

"We can't possibly maintain that floor tile." – Owner's maintenance personnel after the substitute product is installed without their involvement.

What's the point? You have just been advised that the contractor is proposing an alternative or substitute material for what is shown in your construction documents. You have probably been provided with minimal information, perhaps just an email saying *"We can't get material "x" in time for our scheduled start in this area, so we would like to use material "y". Please advise by end of business tomorrow so we can meet our schedule."*

Figure 99 - They both write with black ink – what's the problem?

What are the principles? Ignore for the moment that you have confirmed during the design and documentation stages that specified materials are available. Also ignore the fact that during the bidding process the contractor and his/her subcontractors should have properly confirmed availability, shipping and scheduling of specified materials. Standard construction contracts advise that alternative products from those specified need to be *"...of a quality consistent with those specified and their use acceptable to the Consultant."*[1] or words to that effect. "Acceptable" is not defined, which can work to your advantage. And remember that there is usually no contract between the Architect and Contractor (Design-Build being one exception).

---

1   CCDC 2 2008, GC 3.8.2. Note that CCDC is currently preparing a Guide for Construction Delivery Methods that will be helpful.

Meanwhile, there is nothing explicit in standard client/architect agreements about alternatives and substitutions, or what is acceptable. The closest most standard agreements come is to note that the Consultants need to consider Submittals. And generally submittal review including shop drawing and material sample review are also part of the Architect's basic services during the construction phase – and that includes proposed alternates. Drat!

What are the best practices? In the AAGC universe, alternatives proposed by the contractor or subcontractors should not be casual or willful. There has to be a substantive underlying reason for them because an alternative or substitution is usually a change to the design, AND because of its potential "knock on" effects on your design. Does the new material meet flame spread ratings? Have you exceeded a fire load limit that requires other materials be changed? Have you blown the specified VOC limits for the green building certification (no pun intended)? Is the range of colors and textures in harmony with that finish schedule you sweated over? You need to resolve these issues. They are "issues" because if you get them wrong you will be alone in defending a poorly researched decision, not the contractor.

At the very least the proponent is hoping that a proposed product is *"...of a quality consistent with those specified and their use acceptable to the Consultant."* Surely to goodness it's not up to you to do the legwork around that comparison? Apparently some Contractors think it is. They may initially propose that you should identify the original specifications and research the comparison. Nothing in any form of agreement says you need to do that. The onus is on the contractor, or the subcontractor, to provide the proof of "quality consistent" – and "cheaper" is not that proof, it's not even a criterion!

By this time, my tone of writing probably tells you I find this annoying. You worked hard on that design, they want to change it and they want you to do all the research. Forget it!

But you are a reasonable person, so I have developed two ways of dealing with this issue.

To set the stage for your expectations, your specifications typically include these clauses:

1.4	REVIEW BY CONTRACTOR BEFORE SUBMISSION
A.	Review submittals prior to submission to Consultant. This review represents that necessary requirements have been determined and verified, and that each submittal has been checked and coordinated with requirements of Work and Contract Documents.
B.	Submittals not stamped, signed, dated and identified as to specific project will be returned without being examined and considered rejected.
C.	Action submittals that deviate from the requirements listed in Part 2 below will be returned without being examined and considered rejected.
D.	Information submittals that deviate from the requirements listed in Part 2 below will be returned without being examined and considered rejected.
E.	Submittals for products other than those specified will be returned without being examined and considered rejected, unless the Consultant and Owner have previously agreed that these other products are accepted alternatives to specified products. 1. Notify Consultant in writing at time of submission, identifying deviations from requirements of Contract Documents and stating reasons for deviations. 2. Consultant has sole discretion in accepting or rejecting submittals deviating from requirements of Contract Documents.
F.	Submittals that are incomplete as compared with specified requirements will be returned without being examined and considered rejected.
G.	The time and expense costs to the Consultant of rejecting submittals as described above shall be invoiced to the Owner as additional services, and shall be recoverable in full by the Owner from the Contractor, without deduction or set off. The same procedure shall apply for subconsultants and other Consultants hired directly by the Owner.

Figure 100 - Specifications around Contractor Review

The first part of this set of clauses is fairly straightforward, reminding the contractor of the duty to review before submitting. Clauses B through F are trying to clarify what you will do with incomplete or inappropriate submittals, whether they are for specified items or proposed alternatives or substitutions.

Clause G is the kicker. Because the consultant's contract is with the owner, additional service costs are not recoverable directly from the contractor. Clause G explains clearly how the contractor will pay.

During construction, use an i-Action (i-A) that helps the Contractor to help you. It looks like this:

SUBMITTAL REVIEW:

This form is used to transmit & evaluate a submittal, including a proposed alternative or substitution. To complete this form or reply to this email, select "Reply All", update information & select "Send". "B" = Builder, who indicates: "X" = details attached/ should be attached; "N/A" = not applicable on this occasion; "D" = Designer, who indicates "R" = Reviewed; "RN" = Reviewed as Noted; "RR" = Revise & Resubmit; "X" = Rejected. if you are using an iPhone or iPad you can select a "COMMENT/ SIGN OFFS" text box and dictate comments.

D	B	Type of submittal:	Spec'd	_Alternative _Substitution	Comments
	x	**Indicate reason for proposed alternative or substitution:**		_Cost savings _Schedule improvement _Quality improvement _Other (specify):	
		Description:	Original	Proposal	
	x	Specification section/clause:			
	x	Drawing reference:			
		Submittal information:			
		Manufacturer's literature			
		Installation instructions			
		MSDS data sheets			
		Sample warranty			
		Product sample			
		Shop drawings			
		Other (specify):			

Figure 101 - i-A for Evaluating Alternative or Substitution

Notice that the i-A is entitled "Submittal Review." Yes, use the same template for submittal review regardless whether the submittal is as specified or a proposed alternative/substitution. If there is no proposed alternative/substitution, the contractor simply ignores that column. If there is a proposal, the contractor is required to provide information about the originally specified material/system as well as the proposal.

As with any i-A, you can email it to the Contractor- the instructions are all there. You may choose to delete some information requirements before sending, if they are truly unnecessary. By the way, this is an HTML form, meaning the columns will expand and rebalance layout to minimize overall size as you add text. And if you have a later model smartphone or tablet, content can be dictated.

The recipient (Contractor) needs to fill in all the original specifications, then fill in comparative data for the proposal. That exercise in itself often highlights that the proposal is significantly different from the original, in which case the whole proposal may just disappear.

Assuming the contractor fills in the required information and sends it back, DO NOT immediately perform your review. First have a cursory look – if the proposal may have merit BUT is substantially different than your design, then it is a change to the design, for which you should be paid additional service fees. Forward the i-A to your client, cc'ing the Contractor and noting the Contractor's own research has identified significant changes to the specified design, to advise that you will review this change as an additional service, if the client wishes. Regardless what the Client and Contractor elect to do, when the Client responds your potential costs are now covered.

Some additional advice:

Do not agree to take payment for additional services from the contractor (with whom you have no contract). That would be deemed a conflict of interest at best, a bribe at worst. If you are clear in your invoicing as to what is additional, your client can back charge the contractor at their discretion. This is also important because you have probably had situations where the client has been more than happy to pay additional fees to analyze an alternative because of schedule or budget benefits. If you don't ask you don't get.

Invoice for additional services separately from basic services. That way, if there is a dispute or delay over a relatively small additional services invoice, the main payment will not be held up (by the way, separately invoice reimbursable expenses for the same reason).

When this process has been completed, there are really only two outcomes: 1). The proposal is acceptable; or 2). The proposal is rejected. Either way, you now need to remember to learn from the experience. If you have been managing this in an i-Action (i-A) environment you will now be able to quickly convert the i-A to an i-KnowHow action (i-KH), and tag it with the type(s) of building and construction. If you have discovered your client likes the proposed alternative, you can modify your client preferences so that the next i-WorkPlan (i-WP) you generate for this client includes that preference or permission in the design stages.

Awhile back I had occasion to run into the "we can't maintain this floor tile" scenario mentioned in the second quote at the top. It was on a major commercial project – the client simply forgot to check the substitution with the company charged with maintaining acres of tile.

The learning outcome was a modification to my standard tile checklist to add the following:

Before tile installation begins in any area of the project: Contractor prepare Mockup of each type of tile and grout, including color & layout patterning, acessories like transition strips, and obtain sign-off from Contractor, Owner's representative and tile maintenance company (cleaners). Cleaners should apply cleaning materials to confirm compatibility and performance (e.g., non-slip) Take photos of approved samples, emphasizing patterning, layouts, etc. & attach to mockup reports

Figure 102 - Tile Checklist Modification

RECORD the journey	Consider each proposal as a submittal by the contractor for your review	Summary Principles & Tools	
RESOLVE the issues	Use the Submittal Review content to identify if there is a change to design	i-WorkPlan (i-WP)	Specific procedures for substitutions & alternatives
REVIEW the results	Base decisions on acceptability of alternatives in part on contractor information	i-Actions (i-A)	Submittal review form
REMEMBER & learn to improve	Capture client-specific alternatives for future pre-approval	i-KnowHow (i-KH)	Capture new approvals as i-KH's

LINKS: To look at a Submittal Review i-A set up for alternative or substitution, go to http://bit.ly/aagc33-3

## 34

Submittals -The Basics

"Why do we need to confirm how to build the building we designed, detailed and specified – isn't that the builder's responsibility?" – Young architect to colleague

"If you haven't figured out the detail yet, don't worry. We'll see what the contractor comes up with, then mark it up." – Senior architect to junior

What's the point? One prominent architect in my city (there may be others, I've just not run into them) is infamous amongst the contracting community for preparing construction documents wherein most of the challenging details, corner conditions, odd junctions, etc., are not detailed at all. Hence the second quote above (imagined on my part, but I'm sure it has been uttered). What's particularly annoying is that his design style is defined by the complex conditions that he and his staff refuse to detail.

Although submittals are described in each design team's documents, it is the builder's responsibility to manage the submittal process. BUT it is the designer's responsibility to provide enough design information so that the builder's shop drawings are just that – detailed drawings based on the design drawings enabling the fabrication shop to make the elements designed.

Submittals act to confirm that the actual materials and systems that the builder is about to order and purchase are what the designers want and expect. Remember that you may have specified three or four "approved products", each of which may have slight variations that need shop drawings or other submittals to explain them in the context of the rest of the building. When many of the building's materials have multiple pre-approved options, the permutations are many and submittals help assure the selected combinations work technically and aesthetically. Just imagine: "Window manufacturer type 2's selected products abut cladding supplier's type 3 selected cladding, which incorporates fire exit door supplier's type C door, which has door supplier type H's hardware."

It's very much akin to ordering online, where the good vendor tries to show you the most exact information about what you will be ordering before you place the order, to reduce returns. Returning a shirt is one thing – returning an elevation's worth of curtain wall is quite another, although the evaluation processes are quite similar – size, shape, color, etc.

Figure 103 - Hard to Return Once you've Bought it

Using the same ordering on line analogy, imagine if the suit pictured in an online catalog arrived with different buttons and a different number of buttons (because the best deal to-day was a weaker button requiring one extra), a different pleat detail (based on the button detailing) and in a slightly different color than the catalog, all because you (the designer/purchaser) had neglected to resolve any of these details before placing the order. You would not put up with it in a suit. Yet some designers expect builders to design the details and then draw them, all with the expectation that the designer will then use the shop drawing and submittal review process to either finalize the design, select from the best proffered option, or redesign "on the fly."

What are the principles? For reasons like those noted above, most builders focus significant attention on submittals, knowing that they can adversely affect the construction schedule if resubmittals are required. The best builders first question incomplete or missing details through the Request for Information (RFI) process ("Design it now or design it later." - more about that in Tale #19), then include submittals as well as time to resolve RFI'd details in their schedule. In this way, even allowing for reasonable review times by the designers, submittal review will not delay the overall construction schedule.

Where a builder does not schedule submittals, this is not an excuse for pressuring the consultant for quick review; in fact, it is worthwhile where you see a schedule devoid of submittals to remind the builder that you will require schedule time to review submittals regardless if it has been included in the builder's schedule. It works both ways.

As noted above, proactive builders will review the construction documents carefully and prepare a list of all the submittals they expect to provide, including approximate submission dates and a reasonable time allowance for review. Due to the scale of some larger projects, the builder may assemble submittals into "packages" of submittals arising from the same specification section. Designers should encourage this as it improves the likelihood that related submittals will arrive together.

Title	Package	Total Items
General Requirements	010000	3
Demolition	020500	23
Fibre Reinforced Polymer	032400	3
Concrete	033100	23
Unit Masonry	042000	5
Reinforced Stone Cladding	044200	27
Structural Steel	051000	134
Stainless Steel Railing Systems	057313	5
Rough Carpentry	061000	1
Architectural Woodwork	064000	6
Spray Foam Insul and Fireproofing	072000	6
Metal Cladding	074000	19

Figure 104 - Typical Submittal Packages - 259 items & Counting

Notice in the contractor's submittal list above that there are 27 submittals associated with stone cladding and 134 for structural steel!

The design team is not obligated to participate in the scheduling of submittals, but is obligated to review them in a timely fashion. Timely is a two-edged sword. It eliminates sympathy for the builder who submits 300 curtain wall shop drawings five days before fabrication is scheduled to start. But it also means that if the builder submits those drawings a month before fabrication starts, and advises a reasonable due date at the time, then the designer has little excuse to miss that deadline.

What are the best practices? Regardless if the builder has scheduled submittals or not, they should be an agenda item (or a series of i-Actions) at each site meeting. Here the builder should give the design team advance warning of submittals that will be arriving prior to the next meeting. Some time will be spent reviewing the status of already issued submittals. Reviewers should be prepared to talk reasonably about any delays they are experiencing. The fewer such delays, the fewer are the possibilities of delay claims by the builder.

Package / Submittal / Revision	Description	Required from Trade	Planned Approval Date	Received from Trade	Sent to Consultant	Required from Consultant	Received from Consultant	Returned to Trade	Status	Child Job(s) (if applicable)
020500	**Demolition**									
020500-023	Plaza Level Load for Debris Removal	May 08, 2014	May 15, 2014	Pacific Blasting & Demolition Ltd					Reviewed	
001	Plaza Level Load for Debris Removal			May 08, 2014	May 08, 2014	May 15, 2014	May 09, 2014	May 09, 2014	Reviewed	
032400	**Fibre Reinforced Polymer**									
032400-001r2	FRP-Beam and Column Strengthening			Fibrewrap Construction Canada					Reviewed As Noted	
001	FRP-Beam Flexure and Shear Upgrade Please see the attached product data sheets and associated shop drawings for the FRP Beam flexure and shear upgrade.			Dec 17, 2012	Dec 17, 2012	Jan 04, 2013	Dec 21, 2012	Jan 02, 2013	Reviewed As Noted	
002	Revised FRP Beam/Column strengthening shop drawings including comments from previous submittal as well as column details and floor plans.			Jan 25, 2013	Jan 29, 2013	Feb 06, 2013	Feb 20, 2013	Feb 20, 2013	Reviewed Noted Resubmit	
003	FRP - Beam and Column Strengthening Updated FRP Beam/Column strengthening shop drawings for TP#2 IFC.			Mar 07, 2013	Mar 08, 2013	Mar 15, 2013	Mar 12, 2013	Mar 12, 2013	Reviewed As Noted	

Figure 105 – Typical Submittal Report Details

There are a variety of ways for contractors to manage submittals. The crudest approach is to wrap a transmittal around submittal documents or samples and hand them to the designers. More sophisticated builders will use various document tracking techniques ranging from an Excel spreadsheet to a document management system that may or may not include a simple workflow ("Builder to architect to engineer, back through consultant to builder", etc.).

Drawings and specifications outline a schematic arrangement of materials and products, referring to specifications and typical details. Designers might think their designs are detailed, but they are not. Even the best documents usually exclude some subtle materials, fasteners, etc. Drawings will detail the more typical conditions; submittals such as the contractor's manufacturer's shop drawings will detail every occurrence, even the "one off's." However, as noted above (bears repeating), designers should not expect builders to complete their designs.

Until recently, shop drawings, one class of submittals, often consisted of construction elements like cladding systems or windows floating in space, i.e., with no indication of adjacent construction or materials that might be supplied by others, such as supporting structures. The quality of the context of shop drawings has generally improved and there is no need to accept such "floaters."

B. SHOP DRAWINGS
1. Submit shop drawings in accordance with Section 01300.
2. Indicate materials and profiles and provide full-size, scaled details of components for each type of door and frame. Indicate:
   a. Interior trim and exterior junctions with adjacent construction.
   b. Junctions between combination units.
   c. Elevations of units.
   d. Core thicknesses of components.
   e. Type and location of exposed finishes, method of anchorage, number of anchors, supports, reinforcement, and accessories.
   f. Location of caulking.
   g. Each type of door system including location.
   h. Arrangement of hardware and required clearances.
3. Submit catalogue details for each type of door and frame illustrating profiles, dimensions and methods of assembly.

Figure 106 - Sample Submittal Specification – no "Floaters" Allowed

Where a designer has specified what is to be included in a submittal, such as in the figure above, when the submittal arrives, the specified list can be used as a checklist for the designer. The specification is project-specific, not generic, and if information required is missing, this can constitute grounds for rejection of the submittal. If you are rejecting a submittal on this basis, it is important that you communicate this quickly to the builder, noting what is missing, so there are no grounds for the builder claiming delay. After you are forced to reject a few submittals because they are incomplete, the builder will get the message and start checking submittals for completeness before tossing them your way.

The designer should also check against the schedule as well as the submittal transmittal to see if the scheduled review time is sufficient, and if no review time is specified, what works for the designer? You can be guaranteed that at the next site meeting the new submittal will be on the agenda and you will need to be ready to speak to completeness, reviewability and review time frame.

Samples are another type of submittal that sometimes gets short shrift. Ask for what you need to make a judgment – then be prepared to judge.

C. SAMPLES
1. Submit samples in accordance with Section 01300.
2. Submit one 300 x 300 mm corner sample of each type door and frame.
3. Submit sample showing glazing detail, reinforcement, finish and location of manufacturer's nameplates.
4. Frame sample to show glazing stop, door stop, jointing detail, finish..
5. Manufacturer's Instructions:
   a. Submit manufacturer's installation instructions.
6. Manufacturers' Field Reports: Submit two copies of manufacturers field reports.

Figure 107 - Sample Submittal Specification - Samples

There are occasions when some aspect of the specified sample cannot be easily met. For example, the specification above calls for a 300x300mm (12" x 12") sample. I would probably accept a slightly smaller sample that might be the manufacturer's standard so long as it clearly illustrates the items mentioned in the specification. On the other hand, understanding the corner condition mentioned above is essential, so I would not accept separate pieces of sill, jamb or head sections with the explanation "it all comes together in the corners."

Generally, the biggest challenges with submittals are: a). Defining what you want and need (specifications); b). Getting them submitted in appropriate form/packages from the contractor; and c). Tracking them while they wend their way between the various approving parties (contractor). This is a complexity tailor made for the i-Action:

SUBMITTAL REVIEW:

This form is used to transmit & evaluate a submittal, including a proposed alternative or substitution. To complete this form or reply to this email, select "Reply All", update information & select "Send". "B" = Builder, who indicates: "X" = details attached/ should be attached; "N/A" = not applicable on this occasion; "D" = Designer, who indicates "R" = Reviewed; "RN" = Reviewed as Noted; "RR" = Revise & Resubmit; "X" = Rejected. if you are using an iPhone or iPad you can select a "COMMENT/ SIGN OFFS" text box and dictate comments.

D	B	Type of submittal:	Spec'd	_Alternative _Substitution	Comments
	x	Indicate reason for proposed alternative or substitution:		_Cost savings _Schedule improvement _Quality improvement _Other (specify):	
		Description:	Original	Proposal	
	x	Specification section/clause:			
	x	Drawing reference:			
		Submittal information:			
		Manufacturer's literature			
		Installation instructions			
		MSDS data sheets			
		Sample warranty			
		Product sample			
		Shop drawings			
		Other (specify):			
		Review proposal for:			
	x	clearance dimensions between products and surrounding enclosure elements			
	x	"n.i.c." in someone else's contract.			
	x	any qualifications on the submittal			
		Other:			

Figure 108 - The Essential Submittal Review – i-A format

Being lazy, I have created an omnibus SUBMITTAL REVIEW i-A (see above) that covers all usual submittal requirements, including proposals for alternatives and substitutions. I can put an 'x' in the left Builder column (Req'd) and write whatever notes I need beside the Summary requirement, or paste in the specification for the submittal.

Because this i-A is an HTML document, if I or the recipient want to write or paste in a paragraph of information beside a Summary category, the document will resize itself to fit whatever screen it is being viewed on. As an HTML document it also comes through in email format exactly as pictured below, and the recipient can add a response, including attachments, and return it or send it on its way. Where there are due dates, reminders will be issued.

From: brian.palmquist@me.com
Date: May 25, 2014 at 4:43:12 PM PDT
To: brian.palmquist@me.com
Subject: 00 Material Sample Review Report - |eT1L7p8fCYo=|
Reply-To: brian.palmquist@me.com, actions@quality-works.net

Project: 140327 - Sample architectural project

To reply or add further information, either login to Quality-Works.net, or just hit the 'Reply' button, then add notes in the appropriate table cell and 'Send.'

MATERIAL SAMPLE REVIEW - [Subject]

Req'd	SUMMARY:	_ Reviewed _ Reviewed as Noted _ Revise and Resubmit _ Rejected _ Stored
	ITEM - indicate which are provided, attach as appropriate	Reviewer comments below
	Description of material sample:	
	Other (specify):	
	Applicable specifications/ drawings:	
	Does sample match specifications: Yes / No Comments: Note: Check that sample meets referenced standards as well as specifications. If there is a discrepancy, refer to Prime Consultant via RFI for written resolution	
	any qualifications on the sample submittal?:	
	Other (specify):	

Instructions for use:

- Fill in the form and include supporting information
- Circulate to appropriate Contractor staff and Consultant(s) for review

Action # : A00034
Action Type: Submittal
Sub Type: for review
Priority: 11 Reference
Now Responsible: Brian Palmquist - Copyright 2014 Brian Palmquist - President
Referred By: Brian Palmquist - Copyright 2014 Brian Palmquist - President
Sent By: Brian Palmquist, President, Copyright 2014 Brian Palmquist
Opened: 2014/02/12
Due Date: 2014/05/21
Element: 07 Exterior walls general

Figure 109 - Material Sample i-A received via Email

It is always recommended to place the key conclusions at the top, i.e., "Reviewed", "Rejected", etc. Human nature being what it is, if the contractor reads "Reviewed", the Contractor will return the submittal to the supplier without further review. The supplier may or may not review anything below but will proceed with fabrication and the submittal will be "closed" by the builder.

In the AAGC world, submittals are embedded in a i-WP so are aggregated in the appropriate project phase, floor or building:

Procedure Notes:
2011/06/27 21:23 - Brian Palmquist - DQM collected correspondence re submittals
2010/07/23 9:03 - Brian Palmquist recommended Mario attend next smoke seal turn over reviews as audit
2009/09/02 12:58 - OAC meeeting - reqts for mill test certificates for N.American steel under discussion
2009/09/02 12:31 - Mario asked to check Placecrete on Plan Room
2009/03/23 14:17 - Brian Palmquist - DQA added Task

Actions added for this procedure
Create New Action

Number	Item Name	Due Date	Created	Closed	Floor	Status	Priority	Type - Subtype
A08530 R01	Plaza Design and Schedule	2011/08/10	2011/06/27			Open	05 this month	Information - for Review
A05812 R0	Advisory-Application of Sil-Cure J13		2010/07/23			Open	01 highest	Information - current as of date
A02273 R0	Submitting reports to QA dept.	2009/06/08	2009/06/03			Closed	03 this week	Preferred Practice - superseded
A02188 R0	Steel fabrication in China		2009/05/21			Closed	04 by next meeting or Due Date	Submittal -
A00928 R0	100MPa non-shrink grout submittal	2008/10/29	2008/10/22			Closed	03 this week	99 -

Figure 110 - Submittals Clustered with Procedure (some names omitted)

These submittals can be managed, i.e., added, opened, edited and closed from this location, or from filtered i-A lists such as "open Steel Submittals assigned to Bob".

RECORD the journey	An i-A for each submittal/ set with scheduled dates where known	**Summary Principles & Tools**	
RESOLVE the issues	i-A captures associated correspondence and review process	**i-WorkPlan (i-WP)**	Specific submittal procedure, custom procedures for select clients/ consultants
REVIEW the results	Status indicates what is done and what remains	**i-Actions (i-A)**	One i-A for all Submittals + alternatives & substitutions
REMEMBER & learn to improve	Add client, consultant favorites to i-WP procedures and/or i-KH library	**i-KnowHow (i-KH)**	i-KH's for new alternatives & substitutions

LINKS: Go to http://bit.ly/aagc34-2 for a submittal review form; for a material sample review report try http://bit.ly/aagc34-3

## 35

Submittals - Stick-on Submittal Review

"The paperwork gets separated from the submittals so we never know on site what has been approved." – Superintendent to Consultant

What's the point? One of the contractor's challenges - building on a construction site while managing from an office - is the managing of submittals that are tracked in the office but delivered to the site – range hoods, window sections, ducts, you name it.

What is the principle? When I was a building envelope consultant I got to review just about everything that penetrated the building envelope, including all the mechanical and electrical apparatus. Frequently, submittals I knew I had approved never arrived on site, or somehow the wrong item was indicated to the site as "approved."

What is the best practice? I got tired of this, so devised a very low-tech system that superintendents loved. They would assemble a largish group of envelope-penetrating sample pieces and I would set aside an hour or so on site to review them. I would arrive with the following sticker pre-printed on Ames package labels:

File _____ -8   **Submittal Review**

© 2003 ECO-*design*.ca Architecture + Building Ecology Ltd. This review is for the sole purpose of reviewing general conformance to:

_____Reviewed

_____Reviewed as Noted

_____Revise & Resubmit   _____ the design. concept where ECO-design.ca are Architect Registered Professionals

_____Rejected   _____ applicable Part 5 building regulations where ECO-design.ca are Building Envelope Professionals.

Reviewed by: _____

Date: _____   _____ applicable Part 3 building regulations where ECO-design.ca are Certified Professionals.

E⊂O-*design*.ca

Architecture + Building Ecology Ltd.

3696 West 8th Ave.
Vancouver BC CANADA V6R1Z1
TEL: 604-734-9612
FAX: 604-730-9624

Principals:
Brian Sim
MAIBC FRAIC HON.AIA HON.FCARM
Brian Palmquist
MAIBC MRAIC BEP CP LEED™

This review does not mean we approve or warrant the detail design of other parties or the design inherent in shop drawings, manufacturer's literature or material and system samples. It does not relieve other Designers, Builders, Suppliers and Installers of their responsibility for errors or omissions in drawings, specifications, shop drawings or interpretations of manufacturer's literature or of their responsibility for meeting all requirements of the Contract Documents and applicable building regulations. The Contractor is responsible to confirm and correlate dimensions, for information that pertains solely to fabrication processes and for techniques of construction and installation and the work of all subtrades. E&OE. This review is void if this sticker has been torn, tampered with, removed or altered in any way. (revised 030303)

Figure 111 - Ames Label Size Submittal Review

As I reviewed each item, I would fill in the blanks above, affix the sticker and photograph it for my records and for the relevant consultants including the architect. The superintendent was responsible to convey the results to the project manager (in the office) and to the suppliers. They loved it.

RECORD the journey	Affix a review label to each field sample	Summary Principles & Tools
RESOLVE the issues	Identify review status on the label	i-WorkPlan Specific procedure (i-WP)
REVIEW the results	Photograph the results	i-Actions (i-A)
REMEMBER & learn to improve	Add new approved products to specs	i-KnowHow Capture items not (i-KH) approved

LINKS: To access the form "Alternative or Substitution", go to http://bit.ly/aagc35-2 . A sample has already been added as "Actions added for this procedure." You can download another from the blue http link at the bottom left of the screen.

36

Submittal Breadcrumbs
Critical New or Foreign Products

"So the curtain wall is manufactured in Asia and shipped all the way from there to our site ten miles away. They say it meets spec, that's all I care about."– Architect to Client or Client to Architect

"We have never used this material in our practice, how should we evaluate it?"
– Project Architect to Principal Architect

What's the point? Increasingly, key building envelope and structural materials such as curtain wall and structural steel are being manufactured abroad, in jurisdictions not subject to the laws or standards of the place where the building is being erected. Sadly, it is possible to purchase fraudulent certifications and "approvals" rather like jewellery – adornments to be affixed to products to make them seem to be prettier than they really are.

In the AAGC world **Critical foreign products** are products substantially manufactured in a Foreign Country, whose failure might cause premature failure of the building envelope, other building components or threaten safety and security of occupants. I define **Foreign Country** as a country other than the country where the project is being built. (Yes that means if the project is in Central America and the product is made in the USA, it is from a foreign country)

I define **Critical new products** as products new to your company, whose failure might cause premature failure of the building envelope, other building components or threaten safety and security of occupants. Examples include some of the innovative green buildings materials and systems that are emerging in the marketplace.

What are the principles & best practices? AAGC considers both of these somewhat different categories together because AAGC's recommended process for dealing with them is identical - hence the acronym "CNFP". I apologize that this is a lengthy process, but we are still at the beginning of this global journey. In summary:

First, determine if you have any critical new or foreign products proposed for the project, either in your design or as proposed by the contractor through the submittal, value engineering or alternative/substitution processes:

To determine if the project has a Critical New or Foreign Product (CNFP)	Details
Obtain a copy of the applicable construction documents, including drawings, schedules, details and specifications.	
Identify if these documents include any *foreign country* standards. If so, highlight these for detailed review by Consultants as noted below.	
Identify specific subcontractor qualifications identified in the construction documents. Note that submittals need to address every qualification specified.	
Critical Foreign Product: see definitions above and examples with the Procedure of which this Action is a part	
List at right any countries other than the place of the work where **key raw materials** for the product come from:	
List at right any countries other than the place of the work where the products **main components were manufactured**:	
List at right any countries other than the place of the work where the **product was preassembled**:	
Critical New Product: see definitions above and examples with the Procedure of which this Action is a part	
List at right any product that is a CNP as defined above	

Figure 112 – CNFP Step 1 - do you have CNFP?

Second, obtain the manufacturer's quality control plan and use it to help evaluate submittals. Make sure you read the quality control plan carefully. The first time I was faced with CNFP my nervousness was not alleviated when I realized the quality manual was the worst paste job I had ever seen – completely nonsensical (like some specifications mentioned in a previous tale).

Note that many of the figures that follow are directions for the contractor to obtain or verify certain things. Even though the contractor is usually responsible for the supply and installation of products and materials, consultants need to ensure the contractor is doing these things:

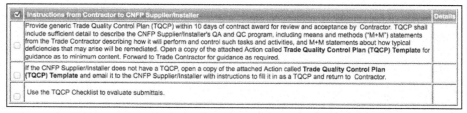

Instructions from Contractor to CNFP Supplier/Installer	Details
Provide generic Trade Quality Control Plan (TQCP) within 10 days of contract award for review and acceptance by Contractor. TQCP shall include sufficient detail to describe the CNFP Supplier/Installer's QA and QC program, including means and methods ("M+M") statements from the Trade Contractor describing how it will perform and control such tasks and activities, and M+M statements about how typical deficiencies that may arise will be remediated. Open a copy of the attached Action called **Trade Quality Control Plan (TQCP) Template** for guidance as to minimum content. Forward to Trade Contractor for guidance as required.	
If the CNFP Supplier/Installer does not have a TQCP, open a copy of the attached Action called **Trade Quality Control Plan (TQCP) Template** and email it to the CNFP Supplier/Installer with instructions to fill it in as a TQCP and return to Contractor.	
Use the TQCP Checklist to evaluate submittals.	

Figure 113 - CNFP Step 2 - Obtain quality control manual

Next, use the submittal, mockup and warranty procedures to get comfort. Notice below the reference to verification by *"qualified registered professionals in the country of the construction, or a Canadian or USA registered professional whose qualifications have been accepted by the contractor, consultants and client."* This sort of wording should always be included in your specifications, since CNFP may be proposed regardless if it is part of your design.

It's essential to know where the buck stops and that it is not in your lap alone. It is equally incumbent on the contractor to be comfortable about these matters. My current employer's general practice is to ensure that the installers of CNFP are local subcontractors with history, credibility and financial strength, so that they can be relied upon if things fall apart (literally as well as legally):

☑	Instructions from Contractor to CNFP Supplier/Installer	Details
☐	Accompany CNFP Submittals with clear written confirmation that the CNFP satisfies all specified standards, in the form of a report signed and sealed by a qualified registered professional in the country of the <u>construction</u>, or a Canadian or USA registered professional whose qualifications have been accepted by the Consultants and the Client.	
☐	Clearly identify on CNFP shop drawings, in English, all materials, sizes, thicknesses, conditions, etc. Bilingual notations are acceptable.	
☐	Test CNFP mock-ups for the purposes of testing to the standards defined in the specifications as well as to the standards of the building codes of the country of the construction.	
☐	Perform all Tests using a qualified registered professional in the country of the <u>construction</u>, or a Canadian or USA registered professional whose qualifications have been accepted by the Consultants and the Client. The testing professional(s) shall confirm in a signed/sealed report that the tests have been successfully passed.	
☐	Draft the test program associated with CNFP well in advance of testing and distributed to Contractor and the Consultants for review and acceptance. The program shall explicitly define each test to be performed and in which sequence. Ledcor and the Consultants shall be entitled to witness any or all tests, at their expense unless otherwise specified.	
☐	Include in test reports details of <u>all</u> test failures, including causes of failure, remedies, and any required amendments to contract documents arising from test failures. Contractor will not accept claims for extra costs or project schedule time arising from test failures.	
☐	When CNFP work samples are completed for review by Consultants and/or the Client, have them reviewed and accepted in writing by a qualified registered professional in the country of the <u>construction</u>, or a Canadian or USA registered professional whose qualifications have been accepted by the Consultants and the Client.	
☐	Accompany CNFP Material samples with a letter confirming conformance to the standards defined in the specifications as well as to the standards of the building codes of the country of the instruction signed/sealed by a qualified registered professional in the country of the <u>construction</u>, or a Canadian or USA registered professional whose qualifications have been accepted by the Consultants and the Client.	
☐	Contractor will forward the reports noted above to the appropriate Consultants/Client for their review and acceptance, which is a precondition to the CNFP work proceeding.	
☐	Regardless whether called for in the contract documents, provide MSDS data sheets for each component of the product	
☐	**Sample warranty:** Provide a sample of the exact warranty that will be offered for the CNFP, clearly indicating proposed length and any limitations. Contractor will review for conformance to contract documents, will also submit to Consultants for review.	

Figure 114 - CNFP Step 3 - Verify through the Process

I have qualified professional input requirements in accordance with my personal experience in Canada and the USA, i.e., I am confortable with that professional standard. Other AAGC readers may of course modify to suit their own comfort levels.

Before erection starts, identify the expectations for field review:

☑	Instructions from Contractor to CNFP Supplier/Installer	Details
☐	Identify any milestone reviews and work "hold" points for approval by the Construction Manager, the Owner or the Consultants that the Trade Contractor may require in order to expeditiously complete the Work	
☐	Contractor, the Owner and/or Consultants may reasonably request additional milestone reviews and work "hold" points for their approval in addition to those specified by the CNFP Supplier.	
☐	The CNFP Supplier/Installer shall not be entitled to extra costs or time associated with such reasonable requests.	

Figure 115 - CNFP step 4 - Field review expectations

Language can often be a challenge with foreign products:

☑	Instructions from Contractor to CNFP Supplier/Installer	Details
☐	Include Project-specific procedures for the delivery, receipt and inspection of materials, the storage of materials, the maintenance of site logs (with details of changes to work shown in construction drawings and field changes, including the record of authorization to proceed with such changes), and the maintenance of all QC-related records	
☐	Provide for review and acceptance by Contractor, the Owner and/or Consultants detailed checklists for completion inspection, shipping and receiving inspection procedures in English, and shall	
☐	Provide copies of all completed checklists to Contractor. Where the local language is not English, provide bilingual (local language/English) checklists.	

Figure 116 - CNFP step 5 - Receiving, inspection & storage

The quality and extent of quality control measures varies widely in foreign countries. Best to treat them like a submittal:

☑	Instructions from Contractor to CNFP Supplier/Installer	Details
☐	Prepare, complete and promptly deliver to Contractor's superintendent quality control checklists that are specific to the type of work being performed by the Trade Contractor.	
☐	Submit proposed checklists to Contractor for review and acceptance at a pre-construction kick-off meeting. Checklists will be reviewed for acceptance by Contractor, the Owner and/or Consultants, and their reasonable comments shall be incorporated in final Checklists.	
☐	Checklist for CNFP shall not be generic. The checklists are to include (in summary form) common deficiencies, typical quality measures and Project-specific requirements, as well as a means to identify deficient work and describe the process for its correction.	
☐	Checklists shall be completed by the CNFP Supplier/Installer for each installation, and copies provided in a timely fashion to Contractor.	

Figure 117 - CNFP step 6 - Checklists and Review

Because the products are foreign, post installation records such as warranties, as-built documents, etc., need to be handled even more rigorously than they would for locally sourced materials:

☑	Instructions from Contractor to CNFP Supplier/Installer	Detail
☐	Make regular submissions to the Construction Manager of as-built information in the form of marked-up shop drawings and sketches. Prior to any application for final payment, the Trade Contractor shall submit to the Construction Manager a final set of as-built drawings consisting of re-drafted construction drawings incorporating all field changes and modifications from site instructions, change orders and requests for information.	
☐	Review and accept CNFP as-built drawings in writing by the same qualified registered professionals in the country of the construction, or Canadian or USA registered professionals whose qualifications have been accepted by the Consultants and the Client, as noted above.	
☐	Submit final draft warranty to Contractor for review. Contractor will compare against original draft warranty to ensure consistency, then submit to Consultants and Owner.	

Figure 118 - CNFP Step 7 – Completion & Post construction records

The recurrent threads for CNFP are: selection of contractors and subcontractors with local history, credibility and financial strength; involvement of qualified consultants who are registered in the place of the construction and have satisfied themselves that the products meet the specified standards through testing equivalent to the specified standards; and continuous involvement and approvals by the client, consultants and the contractor.

RECORD the journey	Identify and track each CNFP	**Summary Principles & Tools**	
RESOLVE the issues	Follow each CNFP to completion of review	**i-WorkPlan (i-WP)**	Specific procedure
REVIEW the results	Multiple review points	**i-Actions (i-A)**	A separate i-A for each CNFP
REMEMBER & learn to improve	Capture CNFP to project, client, consultant or country-specific procedures	**i-KnowHow (i-KH)**	Capture CNFP results (positive or negative) to the i-KH library

LINKS: To access the form "Critical New or Foreign Product", go to http://bit.ly/aagc36-1 . To access the review checklist, click on the blue http link under "File Name"

## 37

Submittals
Shop Drawings as Design Drawings

"The manufacturer had some really great ideas that he has included in the shop draw-ings." – Contractor to Architect

What's the point? It sometimes happens that a subcontractor or supplier will have ideas that refine the consultants' design. Sometimes these ideas arise from the lim-itations of the manufacturer's capabilities; other times from potential manufacturing efficiencies. Whatever the cause, what arises from these ideas are "submittals" that may be dramatically different from the design. The basics of submittals are discussed in previous tales.

There are a number of potential issues with this approach:

Firstly, your design is proposed to be changed. The contractor and subcontractor/sup-pliers may be unaware or likely oblivious to the client's program, local planning/build-ing permit requirements, etc. Also, they may not have considered technical issues such as the integration of their submitted system with the balance of the building's design and construction. You remain responsible for all of those.

Secondly, there may be excellent performance experience and expectations underly-ing your design. Certain materials and systems are assembled in your design based on your experience that they have worked together for you in the past both aestheti-cally and practically. The change may disrupt those proven material and system rela-tionships.

Thirdly, the proposed alternative may not, in fact, be "equal" to the original design. One of my favorites for this is exterior doors and windows. I have analyzed proposals for "equal" doors with inappropriate fire ratings, "equal" windows with poorer resistance to water and air penetration, etc. The performance of such elements is usually verified by laboratory testing to an agreed national performance standard such as CSA[1] or ASTM[2]. I have experienced lower standards, the wrong standards, even no standards!

Remembering that it is the contractor who wants you to agree to an alternative, my ap-proach is to attach Tale #33's Alternative or Substitution i-A by return email, advising the contractor to fill in both the original and the proposal, and that I will start my review upon receipt of this information and approval from the client.

---

1   Canadian Standards Association
2   American Society for Testing Materials

RECORD the journey	Evaluate design changes as such	**Summary Principles & Tools**	
RESOLVE the issues	Consider the larger context of design changes	**i-WorkPlan (i-WP)**	Specific procedure
REVIEW the results	Review against regulatory permits	**i-Actions (i-A)**	Standard submittal review format
REMEMBER & learn to improve		**i-KnowHow (i-KH)**	i-KH for each design change

38

Submittals
Do you really want to get sued?

"But this panel is identical to the one you specified!" – Supplier to Architect

What's the point? I had been summoned to the construction site after receiving a copy of the proposal to substitute a different panel type for the Hardie Panel specified. (I was the building envelope consultant on the project, hired by the client at the architect's insistence). I like Hardie Panel because I understand its benefits and shortcomings and know how to detail around them. I was suspicious of any proposal to substitute something different.

The Contractor was very straight with me – *"This product is more cost effective for us. As far as we can tell it's otherwise the same."*

It certainly appeared to be the same – minor variations in surface texture, otherwise apparently the same. I said I would check it out (the Contractor and developer had not done any research other than cost, were a bit miffed I was going to charge to research the product. But the design Architect supported me – I had saved his bacon a few times before).

What are the principles & best practices? However they arise, proposed changes to materials and systems should be subject to at least as rigorous review as original specifications. This includes the review of the proposed materials themselves as well as interface details, permit implications, etc. The timing of such requests (during construction) and the resulting pressure that may be applied are irrelevant – just remember that you have perpetual liability for what you design, including what you permit to be built from your design. Others do not.

In this instance, when I got back to an Internet-connected computer (pre iPad), on a hunch I entered the name of the proposed alternative product with the suffix "litigation." Three sites popped up around class action lawsuits between various homeowner groups and the manufacturers of this specific product to the tune of tens of millions of dollars. I forwarded the links to the Architect, Contractor and Client with the subject line "Are you Sure?" That was the end of the discussion and we proceeded with Hardie Panel. It took so little time to do the "research" that I did not even charge for it.

I can't say why I thought of "litigation". Perhaps because the manufacturer of the substitute panel was actually a well-known company, so why was their price so much lower than the Hardie competition? As far as I could tell from the websites, the problems arose from product limitations similar to Hardie Panel. BUT whereas Hardie is extremely clear in its directions, details and comments about limitations, this product was not. In any event, bullet dodged

RECORD the journey	Details of any/all alternative/substitution proposals	Summary Principles & Tools	
RESOLVE the issues	Track each proposal to resolution	i-WorkPlan (i-WP)	Specific procedure
REVIEW the results	Review to at least the same standards as original specifications	i-Actions (i-A)	Submittal Review
REMEMBER & learn to improve	Capture new approved materials to master specifications	i-KnowHow (i-KH)	Capture proposals NOT approved to i-KH's for future reference

LINKS: To access the form "Submittal Review - Alternative or Substitution", go to http://bit.ly/aagc38-2 and click on the blue http link at the bottom of the screen.

## 39

Submittals: What's in the back of the Truck? - Check the Submittal in the Field

"But these are the nails that fit in my nail gun!" – Carpenter to Architect

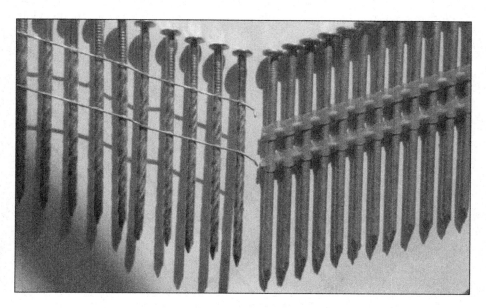

Figure 119 - The Specified Product is on the Right

What's the point? In the wetter parts of North America where I live, we have for some time specified in wood frame construction that nails that might be exposed to moisture should be "hot dipped galvanized" (upper right). Not electro-galvanized (upper left), which deposits only a very thin protective coating, but fully dipped, which provides far greater protection.

Some time ago I also discovered that gun nails (those installed with a nail gun) whose strips are held together with a wire (above, left) will deteriorate quicker than those held with paper or plastic (above, right). The welding of the wire to the nails weakens them to the point where you can often open up a 1-year old nail connection and find a nail from a two-wire strip like the one upper left in the photo broken into three parts, one at each wire link location.

In response, I have for some time specified hot dip galvanized nails with paper or plastic strips:

7.	For fasteners at decks: mechanically galvanized screws, Quik Drive WSCG strip screws treated to an N2000 std, http://www.quikdrive.com/corrosion.htm or pre-approved alternative.Power-Driven Fasteners:
a.	Gun nails:
	1) As specified by structural engineer
	2) Where galvanized, plastic or paper collated, NOT wire collated

Figure 120 – My Gun Nail Specifications

So imagine my surprise during a routine field review to discover that the carpenters were using the dead wrong nail – electro galvanized with wire strips. When I asked why they were using the wrong nails, they responded with the quote at the top of this Tale.

What are the principles & best practices? There was a submittal requested by me for this product – but I did not insist upon it. My bad.

1.7 SUBMITTALS						
A. As per the attached schedule						
List of Required Submittals '+'= req'd,						
NOTE: Each submittal will be considered via a project specific submittal review						
Spec section/ item	Mockup req'd	Manu. Lit.	Shop drawings	Test data	Sample	Warranty
06100 Preservative treated lumber & plywood		+1				
06100 fasteners		+2	+		+3	

Figure 121 - List of Required Rough Carpentry Submittals in Specifications

Worse yet, I had been very specific about the collation method, even adding a footnote to remind me of my own requirements (the footnote appeared at the bottom of the specifications page):

Spec section/ item	Mockup req'd	Manu. Lit.	Shop drawings	Test data	Sa	Sample means, for power driven fasteners, a sufficiently long coil strip to determine the method of collation. NOTE, wire collated are not satisfacto-ry, only plastic or paper collated.
06100 Preservative treated lumber & plywood		+1				
06100 fasteners		+2	+		+3	

Figure 122 - Footnote in Specification - Reminder about Collation

The footnote reads: *"Sample means, for power driven fasteners, a sufficiently long coil strip to determine the method of collation. NOTE, wire collated are not satisfactory, only plastic or paper collated."*

In fairness to myself, I am not sure any amount of pre-planning could have prevent-ed this issue. This was a project where every specification was challenged, usually without warning, among many other problems. I was not rehired for the subsequent phases of the project, and was not at all unhappy about that outcome.

However, the fact that I found this problem underlines the importance of the **REVIEW the results** principle. The occasion when I found this problem was an early field re-view during the framing stage, and the solution I evolved with the builder was to simply double the number of nails in the affected members, with the second set being the specified nails, so that the right number of the right fasteners was in place. Fortunate-ly, the areas in question were covered with cladding, which concealed the machine gun-like appearance of double nailing.

As seen above, it was helpful to embed specific submittal requirements in a simple-to-use format right in the specification. I try to cover all the bases by focusing on submit-tals that propose substitutions first in Division 01 of the specifications:

C. A proposed substitution will be evaluated as follows:
   1. The submitted information to the Consultant is returned, marked "REVIEWED."
      a. This signifies that the proposed substitution, as described in the submittals, complies in the Consultant's opinion with the green building requirements and the other requirements of the project.
   2. Where submitted information is returned marked "REVIEWED AS NOTED", then the Contractor undertakes to assure that the notations by the Consultant are implemented.
   3. Where information is returned marked "REVISE AND RESUBMIT", then the Contractor undertakes to complete the requested revisions and to resubmit the revised information for Consultant review.
   4. Where information is returned marked "REJECTED", then the Contractor shall abandon the proposed substitution, or obtain new, compliant product and recommence the review process.

Figure 123 - Division 01 Specifications for Substitutions

Because I know that many contractors disassemble specifications for distribution to subcontractors and suppliers regardless what designers say, and because I know many subcontractors and suppliers only read the specification sections that apply to their work, I have as much as possible embedded trade-specific requirements for mockups, submittals, warranties, etc., where the trade would typically expect to find their information, i.e., in that trade's specification sections. And because most specifications appear as reams of text, I have also placed these requirements in table form so they pop out of the specification pages. It didn't help in this instance, but usually does.

Table 1 - Type of Submittal Spec Division or section/ item	Mock up req'd	MSDS[2]	REAP Submittal[3]	Green Bldg. Product Information Submit-	Manu. literature	Shop drawing	Test data	Sample	Warranty[5]
05050 steel primer					2	3			
05100 structural steel				2[20]		2			
05400 cold formed metal framing	x			2[22]	2	2	2		
05500 metal fabrications				2[23]		2			
05500 handrails/ guardrails		✦				2			
05500 Bicycle storage racks			SS M3.1[24]						
06100 Dimension lumber[25]			MR 1.3 MR 3.1[26]						
06100 Plywood			MR 3.2[27]						
06100 preservative treatment				2[28]	2		2		

Figure 124 - Tabular Summary of Submittals in Specifications

If a contractor does wish to change specified materials, my detailed instructions look like this:

Procedure Name: Alternatives or substitutions affecting quality

Category: CONTRACTOR

Detailed Description:
Summary: Use the attached review form to evaluate proposed alternatives or substitutions that may affect quality
Purpose: Where an alternative material or product is proposed, to determine whether the proposal imacts the quality of the project.
Includes: Proposed alternatives oir substitutions that may affect quality
Excludes: Alternatives or substitutions proposed by a Consultant or Client, where they are responsible for evaluating effects on quality

Instructions for using:
Process/Procedure description:

1. Fill in the form entitled "Submittal Review - Alternative or Substitution" and submit for review per procedure described below.
2. **Trade** proposes alternative or substitute to specified material/system.
    1. Proponent to compare proposal with original specification
    2. NOTE, where proposal arises from Client or Consultant, they are responsible for evaluation. Record as a Note or attachment(s) to this Procedure the details of their review/approval.
3. Contractor staff review the information provided on the submitted form for completeness and clarity.
    1. Also review for impacts on cost, schedule, etc.
    2. Obtain any necessary clarifications from the Trade.
    3. Review with quality management staff where additional review is needed
4. If proposal appears acceptable, refer to Owner and Consultant - use attached review form or equivalent.
5. **Consultant and Client** review proposal
    1. Accept, reject or request further information
6. Implement if accepted by Consultant and Client.
    1. Inform Trade

Indicate progress and completion of the steps above by entering Notes to this Procedure plus other evidence requested in the steps above. Mark the Procedure as 'Completed' when all steps have been completed and requested evidence has been attached.

Controlled document - Rev 002 - 101130

Figure 125 - Instructions for Alternatives & Substitutions

Basically, tell me what I specified (so I know that you know) and what you want to change it to, and why, and I will review it after discussion with the client about additional services.

RECORD the journey	Summary of all submittals	**Summary Principles & Tools**	
RESOLVE the issues	Using i-A as tracker	**i-WorkPlan (i-WP)**	Specific procedure
REVIEW the results	Be clear about approval of changes	**i-Actions (i-A)**	Specific forms or templates
REMEMBER & learn to improve	Refine specifications to capture new knowledge	**i-KnowHow (i-KH)**	Use i-KH's to capture rejected changes

LINKS: To review the above instructions in more detail, go to http://bit.ly/aagc39-1 and click on "Info" top right of the dialog box.

## 40

Submittals
The Missing Memo

"We met last December and agreed this glass treatment was not correct. I'm sure someone from your office was there, but I didn't keep notes or minutes....surely you did? And we can't find the sample!" – Subcontractor to Contractor

What's the point? This is not my story but I have enough details to tell it, and it illustrates a comment I sometimes make: "We are very good at doing mockups, very bad about recording them."

In this case, the project was a very tall mixed use tower, the tallest and last of a whole-block development by the same developer. The first towers in the development were clad in a very distinctive glass that the developer loved, but hardly anyone else among the urban design community did – among other characteristics, it was so dark it exaggerated Vancouver's winter gloom. When it came time to complete the development, planning authorities insisted on a different glass type for the final tower. The developer reluctantly agreed, samples were submitted and approved, and construction started.

When the building was about 1/3 clad, folks began to notice the glass looked an awful lot like the glass on the preceding towers. Eventually city hall noticed and work was stopped on the project. There were many recriminations and litigation was imminent – then suddenly everything went quiet.

Another prominent local architect (different from the original architect) was brought into the project. He negotiated a design transition from the questionably colored glass at the base to a different glass on the top of the tower. The tower was completed and occupied without any legal action. What happened?

As it happens, the samples of the glass that was approved for the tower went missing, so there was no way for the planning authorities to prove whether the glass at the base was the correct glass or not.

What are the principles & best practices? it's not enough to have an approval process. You must document the process and keep the evidence. Where the review process involves samples, I would (of course) start with an i-A for each submittal to monitor the progress and resolution of the review. In this tale, clear capture of the debates by any of the parties might have eliminated later recriminations. Previous tales discuss the basics of using i-A's to manage submittals.

Where the review process involves a sample, it is important to label the sample so that when stored with many other samples (could be hundreds), there is some hope of identifying and retrieving it.

Sample Tag

QW ID#	
Sample Name	
Spec #	
Quantity	
Submittal #	
Date received by QA	
Trade	
Consultant	
Controlled document	Rev 01 - 110309

Figure 126 - Sample tag

A previous tale included a sample tag that doubles as the review as well, i.e., "reviewed", "reviewed as noted", etc., with appropriate legal verbiage. Where a project's scale or complexity makes an "all in one" label impractical, consider something like the sample tag above, which includes: a unique ID# that connects the sample to an i-A, which i-A includes all of the review details; "Sample Name" that is what you know it as – the "6400 Arctic White" label that the manufacturer affixed becomes "Level 2 east curtain wall" that you have some hope of understanding months or years later; the balance of the entries are self evident. By the way, this sample tag lives at the bottom of my submittal i-A's, so when you fill out the info, print the i-A and use it or even just the tag part to affix to your sample, you have already connected the tag to the i-A.

Few consultants or contractors want to remain as custodians of samples after a project has been completed. In fact, most contractors will explicitly transfer samples to the client as part of the project closeout process. Consultants should generally do the same, remembering to use the i-A or similar approach as a transmittal in order to automatically create a record of that transmission. Where the record shows you gave the sample to the client, and there is any subsequent custody discussion, you will be in the clear.

RECORD the journey	Associate a submittal review & tag to each sample	**Summary Principles & Tools**	
RESOLVE the issues	Use i-A email aggregation to track review and resolution	**i-WorkPlan (i-WP)**	Specific procedure for sample management
REVIEW the results	Only close an i-A when the review process appears complete	**i-Actions (i-A)**	Specific forms or templates for management and transmission
REMEMBER & learn to improve	Refine specifications based on review results	**i-KnowHow (i-KH)**	Capture rejected samples as i-KH's to assist future projects

LINKS: To review the above instructions in more detail, go to http://bit.ly/aagc39-1 and click on "Info" top right of the dialog box.

## 41

Reviewing Construction
Where's the Report!

"Sometimes it's days between a site visit and when I get the Consultant report. And sometimes the report reads like we're on different projects. I walked around with the guy and what he now says in his report is totally different!"
– Superintendent to Project Manager

"Just tell me what I did wrong to-day - before you leave the site!"
– exasperated Superintendent to Consultant

What's the point? Contractors recognize that there are deficiencies and nonconformances on their project – life's like that. And they fervently want to fix them and move on.

This reality often clashes with the consultant's reality, which is complicated by the liability for design and construction decisions as well as their keen sense of design ownership. Also, consultants trained in a simpler/slower era (read pre-Internet) were used to returning to the office, referring to all of the drawings, specs and related documents, and only THEN making decisions and communicating them to the contractor – many of those consultants remain in practice and carry that approach into the digital age.

Consultants worry about saying the wrong thing on a site, missing even one of the elements they have so carefully integrated in the design. Is there a middle ground? In my experience, yes.

What are the principles? The vast majority of consultant observations on site should be positive, i.e., confirming that what is there is supposed to be what is there. In fact, a good consultant field report will focus on specific aspects of construction and confirm they are in general conformance with the construction documents. This is not for the purpose of patting oneself on the back, rather is designed to show that the consultant was looking at many aspects of construction, most of which were acceptable.

Early in my career, the office I was working for designed a housing co-operative with some minor complexities in the floor plans such as jogs in the demising walls between adjacent suites. After occupancy, one family (only) complained of poor acoustics. Do you think we could find even one field report comment or photograph showing a standard/acceptable installation? Not a one! The project architect had focused solely on deficiencies in his reports and photos. We managed to avoid litigation and the contractor opened up the wall to satisfy the occupants – there were no acoustical issues, they just had sensitive ears. We all learned from that to ensure to include records of acceptable installations together with problem areas.

Generally, this confirmation of acceptability by the consultant matters little to the contractor, who assumes all is good unless you say otherwise. Thus the contractor is looking for the problems and solutions, not the positive confirmations.

So why not simply separate the acceptable construction from the deficiencies and nonconformances? There is no reason not to do this, except where the contractor needs specific approval, such as closing in construction that will henceforward be covered and not visible.

What are the best practices? My suggestion to cover both circumstances is quite straightforward and I have used it with much success - identify the problems in a form that can be left with the contractor on site, but reserve the right to add other observations about progress and acceptability of work via separate formal report.

Wait a minute! It sounds like I just added to the harried consultants' workload! Not really.In the AAGC world we use an i-A form to capture field review comments and issues:

Action								
Now Responsible:			Referred By:					
Item:	Field Review Report		Action Number:	A00141 R0	Priority:	11 Reference		
Type:	Sample		Subtype:		Element:			
Status:	Open		Due:		Opened:	2014/09/25	Closed:	
Identifier:			Originator:					

Field Review Report #

Weather:		
Ref. Dwgs:		Grid Refs:
Report date yymmdd:	Review date yymmdd:	Associated Quality Report #:
Review attended by:	Work reviewed:	

This report includes recommendations for use by Consultants to assist Contractors in the timely completion of their work. It is a summary of construction observations, including deficiencies & non-conformances, made in the field, a fabrication or a testing facility & requiring Contractor action. To complete this report or to reply by email, select "Reply All", update information & select "Send; if you are using a dictation-equipped smartphone or tablet, you can select an "Observations" text box & dictate comments."D" = for action by Designer; "B" = for action by Builder. Use "Images" to identify any attached photos or sketches. "Action By" is for the Builder's assignments. Use the "Verify" column to identify if the Consultant expects verification of completion by "I" = Initial; by "P" = Photo; or by "F" = followup field review.

D	B	#	Observations	Images	Action	Verify

Figure 127 – Field Review Report (Top)

(The form actually has several repetitive rows that have been deleted to save space here.)

This report evolved from an earlier multipart paper version, but the principles are identical.

Note the right hand columns where the consultant is asked how he/she wishes to confirm completion of the issues, via initial, photo or return to site. Think of how many times you have gone to site and seen work you wanted to see already covered; or had your time wasted being called to site to see work you were happy to have had confirmed via photo.

The consultants' potential concern about use or misuse of the report is addressed in the first instruction at the bottom of the form:

1. Email this report to the builder before leaving the site or as otherwise agreed with the Builder. The Builder understands the report may be refined after review by the Designer's office.
2. Uniquely number all reports and report items. Keep the same filename format.
3. Periodically attach an updated cumulative list of deficiencies you consider to be unresolved. Your list of cumulative unresolved deficiencies serves two purposes:
   a. It becomes your checklist when you re-examine previously reported issues.
   b. It signals to the Builder whether a corrective action was satisfactory (without additional correspondence) and allows closure of completed items.
4. In the "Observations" Column, include the location of comment, room #, grid reference, etc. as well as details. Describe each deficiency carefully and unambiguously. Give a precise location of each item with Floor level and gridlines, to allow the Contractor to follow up.
5. Attach photographs at the bottom of this report. Do not insert them into the table cells above as this will distort the appearance of the form on smartphones and tablets.
6. To ensure your reports are acted on expeditiously:
   a. Always use the same email subject line when sending a report to Contractor.
   b. Create a standard distribution list for your reports that includes the Contractor's named email address
7. These are general guidelines. They are not intended to replace the special reports that require update and re-issue, such as concrete lab reports and fireproofing inspection reports.

Figure 128 - Field Review Report Instructions - Look at #1

Most of these instructions are more or less automatically fulfilled by using i-A's, but you should read them as they will improve your communication with the construction site. The essential instructions:

Leave a report before you leave the site – it's okay to indicate that you reserve the right to reissue it after review at your office.

Identify your reports with some sort of numbering so that they can be referred to later, say, in a meeting.

Be clear about how you want evidence of deficiency correction presented – via initial, photo, or is a return visit required?

Periodically issue a list of items that remain unresolved as far as you can tell – there will be occasions when the contractor feels an item has already been dealt with. Better to identify a misunderstanding now than at the end of the project.

This approach also preserves the consultant's ability to supplement this content with a report in a consultant's other preferred format.

The contractor can use this same format to marshal subcontractor forces and get items corrected. The report can simply be forwarded to one or more subcontractors via email. The subs can attend to items mentioned, record their replies, including photographs and reply to the email. After verification, the superintendent can then forward this evidence back to the consultant, for final review and closure. If this is all done from the originating field review report, your i-A should automatically capture each issue, reissue, annotation and photo associated with completing the work, for best records and easiest management.

Of course, this approach can also be used by a contractor for internal review.

On some projects it may be desirable to handle each non-conformance or deficiency as a separate i-A. This is discussed in another tale.

RECORD the journey	A report for each site visit	**Summary Principles & Tools**	
RESOLVE the issues	Issues requiring Contractor action left on site before departing	**i-WorkPlan (i-WP)**	Specific procedure
REVIEW the results	Use originating report to track resolution	**i-Actions (i-A)**	i-A for each report or each issue
REMEMBER & learn to improve	Feed back into i-WP	**i-KnowHow (i-KH)**	

LINKS: To access the form "Instructions for a field report", go to http://bit.ly/aagc41-1 - forms, templates and instructions are at the bottom of the screen.

## 42

Reviewing Construction
Check the Weather

"Have you gentlemen read what it says on the label on the pail about applying this stuff in the rain?" – Building Envelope Consultant to Installers

I had arrived at the site on a rainy day near the end of a residential project; it had been raining all week. As I walked from my car towards the site trailer, I noticed two workers on ladders with long paint rollers, applying something clear to the building's brick walls. Curious about what they were up to in the inclement weather, I approached.

At the base of their ladders was a five-gallon container of transparent water proofing material. In letters about 2 inches (50 mm) high, the instructions advised against applying the material if it had rained within the preceding 24 hours or was forecast for the next 48.

"Gentlemen," I yelled from the ground, *"Have you read the instructions on this stuff? You shouldn't be applying it in this weather."*

Their response was colorful but not printable, and made it clear they were not about to cease and desist.

"Okay, guys," I continued. *"Just so you know I am heading to the superintendent's trailer from here, and I will be rejecting all this work. Just sayin'."*

The workers swore a bit more, then got down off their ladders. The affected areas were recoated some days later when the weather improved.

What's the point? There are a couple of important concepts here. First, it is important to verify the application limitations of the actual materials selected. Do not assume the installers have either read or understood even the clearest instructions. And don't assume they were not directed to proceed in any event, in order to meet some schedule. This does not mean you have to rush out to construction sites every time the weather changes, but it does suggest that if you are on site in changed weather, you need to think about the implications.

To underline this concept, some years ago, my then business partner was acting as an expert in a legal matter, helping defend an architect who had designed a school gymnasium whose east-facing wall leaked badly and required expensive remediation. Such work (preparing expert reports) is often grueling and thankless, so I was surprised to hear a howl of laughter from his office one morning. I investigated and he handed me one of the architect's field review reports, which said in part:

"Weather, -5 deg C (20 deg F), light snow.", then

"...stucco application proceeding well on east elevation of gym."

You don't install stucco in -5 degC weather without suitable protection such as insulated tarps to hold in temporary heat!

Needless to say, my partner's effort thereafter was a rearguard attempt to minimize the damages.

Finally, remember that consultants cannot direct workers – only foremen and superintendents may do that. However, this does not prevent you from advising workers that you plan to take action that will negate work they are in the process of executing. There is no point in allowing inappropriate work to continue any longer than necessary.

What are the best practices? Record the weather each time you visit a site – even better is to note the weather from yesterday and the forecast for tomorrow, especially for weather-sensitive work such as concrete, stucco and masonry. Just the act of writing these down may trigger reminders about limitations to current and proposed work.

Weather: ☐ Sunny ☐ Cloudy ☐ Mixed ☐ Rain showers ☐ Rain ☐ Snow ☐ Other/details:_____				
Day before:	Weather:	Temp.	degC High	degC Low
To-day:	Weather:	Temp.	degC High	degC Low
Day after:	Weather:	Temp.	degC High	degC Low

Figure 129 - Weather info at top of checklist

As part of organizing a site visit, quiz the superintendent before you leave on what work is currently proceeding, not just what you are going to look at. Think for a moment about any weather-related measures the contractor should be considering in relation to that work, such as tarps, temporary heat, dehumidification, etc. Remind him if you think there is a need to do so. Then when you arrive, check to see if the work is protected, and not just the work you have come to review.

Being cautious such as described above does not affect your professional liability. In fact, if there is ever an issue and it becomes clear from your reports that you have been as diligent as a consultant can reasonably expect to be, your defense will be strengthened. And frankly, if the superintendent realizes you will ask questions, then check the answers while you are site, she/he will work to match your diligence.

RECORD the journey	Record the weather on each field review report	**Summary Principles & Tools**	
RESOLVE the issues	Identify issues arising from weather changes	**i-WorkPlan (i-WP)**	Specific procedure
REVIEW the results	Re-review work installed during inclement weather	**i-Actions (i-A)**	i-A for reports and issues
REMEMBER & learn to improve	Feed back into i-WP	**i-KnowHow (i-KH)**	

LINKS: To access procedures and forms related to consultant field review, go to http://bit.ly/aagc42-4 - forms and instructions are at the bottom of the screen.

## 43

Reviewing Construction
Continuous As-Builts

"We set up the drawings and specs so our team could walk around the site and find any details they need on an iPad. Then we linked in RFI's and Submittals. Now the client, the consultants and the building inspector bring their iPads when they visit site and synchronize the latest updates in a few moments while checking in at the trailer. They know they will always have latest/greatest information. And they can mark up observations on their copy of the documents and communicate them without affecting our masters. It cost the equivalent of 2% of my salary during the duration of this project to set up."– Proud Superintendent to impressed Quality Director.

What's the point? Fortunately in the case of the five-gallon pail noted in the preceding tale, the instructions were plain as day and I was familiar with the product. But it's becoming more and more difficult to keep track of all of the products we specify, let alone which of the acceptable suppliers and products has been selected by the contractor for each project.

What's the principle? It is now possible to link construction documents with approved submittals, answered RFI's, etc., place them on your tablet and update them either in real time, or at worst, each time you visit a construction site. Contractors may have access to this same information (in fact, contractors are increasingly setting it up for their own use and may give you access if you ask nicely). By the time this e-book is published I expect almost all contractors will have such a system in place.

What's the best practice? Although there are several types of tablet and associated software, the essential approach works like this:

Start with a tablet computer. It needs Wi-Fi capability but need not have a 3G/4G cellular connection (although I recommend that for other reasons). You will also need a conventional laptop or desktop computer, although all this may be doable from a tablet by the time this e-book is published.

Using PDF markup software such as BlueBeam on the laptop, you or a local print shop can link all of the drawings, details and specifications so that when you tap a reference number on a PDF drawing on a tablet it will take you to the referenced detail or specification.

This becomes interesting when you can also link project documents such as submittals, RFI answers, etc. What you are now building is a continuous as-built record.

During the course of a project, the laptop is used to link new project documents to the construction documents, creating a comprehensive record.

The "magic" happens through use of a file transfer program such as DropBox or Sharefile. Current documents, including updates, are downloaded to onsite tablets whenever it suits their owners. For example, if the contractor is maintaining the records and the architect comes to site once a week, updates can be downloaded in a few moments at the start of each weekly visit.

Construction is a stressful business. Continuous as-builts give team members a bit of needed comfort and assurance in a changeable world.

RECORD the journey	Hyperlink the documents	Summary Principles & Tools
RESOLVE the issues	Capture additions and changes	i-WorkPlan Specific procedure (i-WP)
REVIEW the results	Download to a tablet for "latest/ greatest"	i-Actions (i-A)
REMEMBER & learn to improve	Download the final records to a thumb drive – as-builts	i-KnowHow (i-KH)

LINKS: Bluebeam software can be found at http://bit.ly/aagc43-1

## 44

Reviewing Construction
"I Just do Drawings" - Field Review Scope of Work

"My fee only includes a monthly site visit. I can't come every time you have a problem or can't read the documents." – Architect to superintendent

What's the point? Professionals get to define their extent of field review. Standard industry contracts say things like *"at intervals appropriate to the stage of construction that the Architect, in his or her professional discretion, considers necessary to become familiar with the progress and quality of the Work and to determine that the Work is in general conformity with the Construction Documents, and the reporting thereon."* But this "discretion" is not a license to never show up. The trick is that in the event of a problem, the professional may need to defend that "professional discretion" was appropriately considered.

GENERAL REVIEW / FIELD REVIEW

General Review / Field Review means the visits to the *Place of the Work* (and where applicable, at locations where building components are fabricated for use at the *Project* site), at intervals appropriate to the stage of the construction that the *Consultant*, in his or her professional discretion, considers necessary to become familiar with the progress and quality of the *Work* and to determine that the *Work* is in general conformity with the construction documents.

Figure 130 - One Professional Definition of Field Review[1]

What are the principles? There are so many provinces, states, cities and towns that common practice is impossible to define, but there are some useful guidelines below.

Think about the phrase "...in general conformity..." then ask yourself the question, is that enough for your project? If your project is a big box retail store that is 90% repetitive and more structural than anything else, maybe that means you need not visit often. I am certainly aware of consultants who take that approach. Or perhaps you visit the 90% just enough, and the 10% that distinguishes your design somewhat more?

Awhile back I worked on a "big box" project that was 90% standard box, 10% specific to the project location (I know, not exciting, but bills must be paid!). The 10% "special" required the use of EIFS (Exterior Insulation Finishing System), not my favorite material.

Why am I leery of EIFS? Because some years ago I took a course enabling me to be a certified EIFS inspector. Then I noticed the EIFS installation checklist included 113 items. That's 113 places I need to check that all has been done right. Impossible!

Needless to say, I focused a lot of effort on that 10%!

What are the best practices? For a "typical" building design project, I have identified 15 "touch points" for a typical project's field review phase:

---

1   AIBC Standard Form of Contract 6C between Client and Consultant, Definitions, page 2 of 3, www.aibc.ca

Process Name
Drawing Revisions in Ident
Identify and remedy weather-related deficiencies
Provide additional field/general review services
Coordinate Consultant services
Collect, review and resolve Others' site visit issues
Monitor moisture management strategy
Monitor Mockup construction
Monitor Testing of construction
Prepare list of required field reviews
Provide Field/general review
Review field review reports of others
Monitor moisture management strategy
Perform pre-occupancy field/general review
Identify deficiencies and verify repair
Establish $ holdback for deficiency completion

Figure 131 - 15 Consultant Field Review Responsibilities

If a project's scope is modest, many of the 15 will amount to a "one of" event. For larger projects, many of these, e.g., "Monitor Construction Testing", become major events in themselves and may stretch over many months.

For a larger project, the number of RFI's, changes, clarifications, alternatives and substitutions, non-conformances and deficiencies will be in the hundreds, often thousands of separate items. Each needs to be identified, captured, communicated and resolved, else it will come back someday and require much greater time, effort and cost.

Regardless of project scale we need to make the effort to capture new knowledge and lessons learned "on the fly", in the heat of project execution before the project teams have dispersed to other challenges.

RECORD the journey	Identify your extent of field review	**Summary Principles & Tools**	
RESOLVE the issues	Track all issues arising to resolution & closure	**i-WorkPlan (i-WP)**	Min. 19 touch points
REVIEW the results		**i-Actions (i-A)**	Specific forms or templates
REMEMBER & learn to improve	Capture i-WP refinements "on the fly"	**i-KnowHow (i-KH)**	

## 45

Reviewing Construction
"What do you Mean it Leaks?"

"Of course these doors leak! They have no water resistance rating. Anyone who says different is nuts!" – Door supplier to Architect

"You'll never believe it! After that big rainstorm last night the wood floors inside six of those ten doors buckled. I guess we'll have to do something. What a surprise!" – Owner to Architect

What's the point? I was the building envelope consultant for a row of south facing townhouses exposed to the weather in the rainy city of Vancouver. There was no overhang over the doors. We had specified an appropriately robust water resistance rating for the doors, well in excess of what they appeared to be as installed (you get to know these things after awhile).

A water test had been scheduled for one of these doors as well as for some windows in the project (the windows passed with flying colors). On the day of the test the door installer was on site (always important to assemble the key trades in case there are issues or information is needed). When I started to set up the test apparatus and he became aware of my intentions around testing, he offered the first quotation above. Apparently no one had told him he was to supply water resistant doors!

Figure 132 - Water Test Apparatus (Before Alignment)

Sure enough, the doors leaked from almost every seam even before water pressure was added. No amount of caulking, sealing or flashing could make these doors pass the water test.

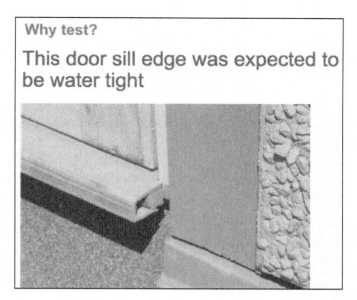

Figure 133 - Door sill edge

And there the matter rested for a week. The developer and builder insisted my requirements (straight from the building code) were overkill. The architect just wanted the problem resolved.

Then it rained. We had one of those frequent wind driven rainstorms in Vancouver, not even a very severe one.

The day after the storm the developer called me and said *"Almost all of the townhouse patio doors leaked and the hardwood floor inside the doors lifted! What can we do?"*

Figure 134 - The Wood Floors Lifted

No mention of my original objections or the failed test – now it was "what can we do?" The solution was actually very simple. The architect designed a very attractive canopy over each door, which provided sufficient shelter to avoid a recurrence of the leakage.

What are the principles? Most building codes establish either directly or through reference to standards a wide variety of performance standards, including water resistance of openings like doors and windows. Scientists and engineers have developed (usually) simple ways to measure whether an installation conforms to the standards.

Just as designers are not expected to review each and every installation, rather a representative sample, so it goes with testing. Codes, referenced standards, (sometimes) insurers or local professional standards will provide guidance about quantity and type of testing.

What are the best practices? Regardless of what codes and standards mandate or recommend, in the end the design professionals will be responsible for identifying the nature and extent of testing. You cannot count on others to record such tests. You should always keep your own records, supplemented by a significant number of photos, even videos in some instances.

After you review the established quantity of installations, you may need to resolve issues arising from testing or mockup details. Finally, where you identify issues, you need to remember and modify your design and/or construction standards to suit.

Most contractors have one or more pre-mobilization meetings with each subcontractor before they start their work scope. Such meetings are usually private between the contractor and subs, but if an event such as described here occurs, it is reasonable cause for the designer to remind the contractor at an OAC meeting (Owner Architect Contractor) that it is the contractor's responsibility to manage the subcontractors in this way.

It is true that the contractor is responsible to identify and provide whatever testing is specified in the construction documents. But what do you do if something is missed or a test fails?

In such circumstances there is usually significant pressure put on the designers to accommodate the failed testing in some way – you will usually be less comfortable than you would like.

The only effective solution to this is to avoid it in the first place. Thorough design review before tendering will help:

Exposed Swing Doors	[New construction: [Vinyl][**Metal**][**Wood**] frame doors.] [No work shown.][**describe**]
	Submittals [will be required demonstrating] [**demonstrate**] [**A3/B3/C3**] [**describe appropriate**] performance.
	Roof overhangs to protect exposed swing doors [required], [not detailed][details [not][appropriate] – [see detailed comments below]]
	Air/ Moisture/ Vapor Barrier Continuity:
	- air barrier continuity at [interface with walls]: [not detailed][details [not][appropriate] – [see detailed comments below]]
	- moisture barrier continuity at [interface with walls/roofs]: [not detailed][details [not][appropriate] – [see detailed comments below]]
	- sub sill membrane with back dam [not detailed][details [not][appropriate] – [see detailed comments below]]
	- vapor barrier continuity at [interface with walls/roofs]: [not detailed][details [not][appropriate] – [see detailed comments below]]
	[**Assemblies [not] appropriate**]

Figure 135 - Design Review Template for Exposed Swing Doors

I have developed a 9-page design review report – this is the "exposed swing door" portion. The [square brackets] indicate design choices, with accompanying items to consider.

For this tale of poor doors, the report looked like this:

Exposed Swing Doors	New construction: Wood frame doors.
	Submittals will be required demonstrating **A3/B3/C3** performance.
	Roof overhangs to protect exposed swing doors not detailed
	Air/ Moisture/ Vapor Barrier Continuity:
	- Air barrier continuity not detailed
	- Moisture barrier continuity at interface with walls/roofs: not detailed
	- Sub sill membrane with back dam not detailed
	- Vapor barrier continuity at interface with walls/roofs: not detailed
	Assemblies may not be appropriate – details noted above required

Figure 136 - Exposed Swing Door Review Comments

I was the building envelope consultant on this project, not the architect. So I could only make recommendations and say, if asked, "We can address these items now or during construction while standing on the site in the rain."

RECORD the journey	What records are needed?	**Summary Principles & Tools**	
RESOLVE the issues	Typical issues arising	**i-WorkPlan (i-WP)**	Specific procedure
REVIEW the results	How do we measure how we have done?	**i-Actions (i-A)**	Specific forms or templates
REMEMBER & learn to improve	How can we do better?	**i-KnowHow (i-KH)**	Knowledge items to look for

LINKS: To access a building envelope design review checklist, login to Quality-Works and go to http://bit.ly/aagc45-1

## 46

Reviewing Construction
Showing up Unannounced

"It's always busy here when I come on my regular Wednesday afternoon site visit. But they don't seem to be making much progress week to week."
– Project Architect to Principal

What's the point? It was not by any means my first project as Principal of the firm, but it was one of the first where I had delegated so much to a junior colleague. I should have noticed the signals earlier.

My junior had fallen into the practice of always visiting the site at the same time. The Contractor knew this, so he staffed up the site for those occasions. As soon as he left the contractor sent the crews to his other jobs.

When my colleague's words finally resonated, we hopped in a car and went to the site – it was not the usual review day. Sure enough, the site was deserted. I contacted the Contractor for an urgent meeting the next morning. He did not show – instead, he declared bankruptcy later that day. He literally skipped town for awhile but was back in business within six months (not on our project).

For a week I sweated bullets while a forensic accountant hired by the client reviewed how much money we had certified for payment to the Contractor. The good news – at the end of the week he advised that his estimate of what should have been certified was only 3% different than mine. (Anything less than 10% is considered a reasonable margin of error.) Whew!

Every once in a while in a professional career, something happens that may not be your fault, but nonetheless makes your blood run cold. This was one of those few occasions in my career – not pleasant!

What should be the timing and frequency of field review?

What are the principles? Most Consultant contracts leave the quantity and timing of field review up to the Consultant. This lack of guidance is unfortunate. It allows knowledgeable clients to pressure Consultants to perform less field review, hence charge less fees. It encourages infrequent review, as little as once per month, even less where the Consultant is not acting as payment certifier.

The link between payment and site visitation is a double-edged sword. Since Contractors bill monthly, the consultant team must visit the site at least monthly in order to evaluate progress – but need it visit more?

For legal and insurance reasons, architectural and engineering associations have declined to advise about frequency of field review for a long time, in case their advice is too closely followed, turns out to be wrong and they are sued. I take a different approach, based somewhat but not only on this tale from the trenches.

It starts with the proposal for a project. The field review effort is not just about showing up! Rather, it includes: preparation time; travel time; time on site; travel time back to the office; reporting time; issue follow-up time. Depending upon the site location, this can easily stretch from a half day to a full day. All of this must factor into your calculation of the fees associated with field review.

My initial fee calculation works this way: What is the duration of the job's construction phase? If I don't know I ask the Client or a friendly contractor. What is the project likely to be, i.e., one building or five, low-rise, mid, or high, etc.? Impossible to know with certainty but since you are the designer, it is often susceptible to a good guess. How many reviews per building, per floor, etc.

For example, for a townhouse building as the architect: one review at foundations for dampproofing/ waterproofing, one at pre-insulation (firestopping, etc., visible), one at pre-board (before drywall is applied, but after insulation and vapor barrier are in place), one or two during finishing, one final (if it's more than one it's extra). What's that, four or five, some of which can maybe be combined with month end payment certification reviews for a smaller, simpler project. There may also be some efficiencies with staggered construction of multi-building sites.

A	C	RA	N/A	Minimum Required Field Review	Comments/ Conclusion/ Action - Minimum Field Review
x			x	Foundations - Includes dampproofing, waterproofing, weather barrier under slab, below grade/slab insulation	Once per building or once per elevation
x	BE		x	Moisture content - wood frame - Requires bldg be enclosed with roofing, weather barrier & all doors/windows in place except at drywall loading locations.(i.e., should be entirely weather protected)	Once/bldg -> townhouses; once/floor others. Note, Contractor identify RA review requirements for framing.
x	BE			Window/Air Tests - Typically 1% of windows are tested, min. 2 per phase, one early and 1 late in phase. The later test may include air testing of a typical completed unit. Coordinate with Consultant which bldg(s) have windows tested.	Divide between phases or buildings; approx. 1/2 earlier, 1/2 later
x	BE M/P E		x	Pre-drywall - Requires insulation in place, and as req'd, ADA sealant and gaskets, window sealant, poly v.b., f'stopping, ductwork sealed, other special conditions, etc.	Once/bldg -> townhouses; once/floor others. Contractor identify RA review requirements for pre-drywall
x	R			Roof - Requires roofing & flashing complete - important to conduct early in project to establish performance requirements.	R = independent roofing inspector if spec'd
x	BE			Exterior Pre-cladding - Includes preparations for cladding on all exterior wall surfaces, i.e., moisture barrier and strapping measures in place. Often combined with other milestone reviews depending on supt. organization of work.	Once/bldg -> townhouses; once/floor others.
x	BE			Exterior Envelope - Includes substantial completion of all exterior surfaces - may be combined with Pre-occupancy if circumstances permit.	Once/bldg -> townhouses; once/floor others
x	M/P E		x	Firestop systems - Submittal required describing each f'stop system proposed, for CP review.	Once/bldg -> townhouses; once/floor others. Contractor identify RA review requirements.
x	M/P E			Chases - fire rating pre-board - For lower fire ratings (up to 1h), typically requires 3 of 4 sides of chases be drywalled and taped/sanded and 4th side be ready for sealing. Verify method of sealing 4th side as acceptable per code/local building authority.	Once/bldg -> townhouses; once/floor others. Contractor identify RA review requirements.
x			x	Fireplaces - fire rating behind, pre-board - Requires wall areas behind fireplaces have insulation, vapour retarder and taped drywall in place.	Once/bldg -> townhouses; once/floor other. Contractor identify RA review requirements.
x	M/P E		x	Corridor ceiling drops - Ceiling areas insulated as required, taped with penetrations firestopped.	Once/bldg -> townhouses; once/floor other. Contractor identify RA review requirements.

Figure 137 - Partial list of my Required Field Reviews

So multiply four or five times the number of buildings and floors times the average cost of a visit, i.e., prep, travel to, on site, travel from, reporting, follow-up. There's your fee. Magic!

What are the best practices? Actually, there is no magic to fees unless you work it too hard. I calculate what the fees are based on, number of visits and cost of a visit, then compare it to recommended fees to ensure I am not way high or way low.

My secret is that I seldom do construction phase work for a fixed or percentage fee. Rather I tell my client exactly what I have calculated and the cost of a visit, and identify that as a fee budget. Then I table the kicker – if the contractor is more efficient and/or has fewer issues, then I will require fewer visits and fees will be reduced; conversely, if my time is abused because of site inefficiencies, my fees will be higher. And I tell the client I promise to raise my hand early and say if I think I am going to go over budget as soon as I sense it, rather than as a nasty surprise.

Construction	Construction contract administration	
	[Enhanced] field review – [based on] [an historic average [[4] hours/unit] for projects of this size range][based on enhanced review [of **detailed and emergent building conditions**] of foundations, moisture content, pre-drywall, cladding, roofing and pre-occupancy][and][[assuming] [one][**four**] review[**s**] per [building][floor][other].]	$

---

[1] Hourly fees: [By law, *[Company]*'s blended hourly fees are based upon the Tariff of Fees published by the [association name]]. Since the tariff is amended periodically, the fees for each Project are stated clearly in each *Contract*. *[Company]*'s blended hourly fees for this project are $[ ] for Principals $[ ] for BEP's & CP's & senior Architects; $[ ] for intermediate staff, building envelope technical and field staff; $[ ] for junior architect/building envelope technician; $[ ] for architectural students and administrative support staff.

Figure 138 - My Proposal Wording Around Field Review Fees

In the figure above, the square brackets indicate places where I need to make choices when I am preparing a contract. There seem to be a lot of [square brackets], but once you have done a couple of these the process is very quick.

1. Construction phase Additional services:
   a. Services resulting from deficient *Workmanship*, materials or products, including the re-review and/or retesting of deficient *Work*;
   b. Additional *field reviews* resulting from decisions by others that require [Company] to attend on construction sites more frequently or for longer periods of time than reasonably scheduled by [Company];
   c. Additional testing and reviewing required as a result of tests, mockups and reviews that indicate noncompliance with requirements;
   d. Review of products other than those originally specified, [including those specified in [Company]'s design reviews];
   e. Review of alternative designs or design changes [not originating from [Company]'s design reviews];
   f. Services for repairs of damage during construction as the result of fire, man-made disaster, or natural disaster, and similar;
   g. Services made necessary by the default of the *Contractor* or Client under the *Contract* for Construction;

Figure 139 - My Contract List of Additional Construction Phase Services

Consultants do not have contracts with Contractors unless they are working for them on a design-build project. But this does not prevent Contractors from abusing the time of Consultants. By establishing on paper a clear listing of anticipated field reviews, and checking against it, I can advise early if I will exceed my budget.

In order to avoid the problem I had with the Contractor who was sending his forces elsewhere, I (now) also drop in unannounced at random intervals during construction, usually before or after another scheduled event in the project's vicinity. Remember, Consultants are in charge of the extent and timing of their field review. Occasionally a Contractor may get annoyed at this practice – tough! Read the construction contract and you will find you can show up pretty much whenever you wish, subject to site safety requirements such as wearing personal protective equipment, safety orientation, etc. Don't make the same Wednesday afternoon mistake as my associate!

RECORD the journey	Calculate the fees based on the scale and scope	Summary Principles & Tools	
RESOLVE the issues	Use budgets rather than fixed fees to manage uncertainties	i-WorkPlan (i-WP)	
REVIEW the results	Check actual time regularly against estimates, & advise the client immediately of any likely overages	i-Actions (i-A)	
REMEMBER & learn to improve	Track your time carefully on every project & feed that experience into your fee calculations	i-KnowHow (i-KH)	Historic costs for various clients, communities & project types (10C's)

LINKS: Go to http://bit.ly/aagc46-1 for my complete list of minimum required field reviews.

## 47

Reviewing Construction
Inadvertent Damage & the Value of Reinspection

"I can't install the flashing properly unless I trim back the roofing."
– Installer to Superintendent

What's the point? The developer, architect, roofer and contractor had just completed a successful review of the large, low slope roof on the condominium project, and cleared the flashing contractor to install the perimeter flashing.

As they reached ground level, the developer realized he had left his clipboard on the roof (pre-tablet era), so climbed back up to get it. Next we received a phone call from the roof (post cell phone arrival). "You've got to come up here, you won't believe this." So we all trekked up four floors of exit stairs and a temporary ladder to the roof, where we saw the developer in discussion with the flashing installer.

Imagine our shock as we approached the two and saw a neat strip of roofing material lying loose atop the installed roofing – almost all of the roofing material atop the short parapet had been neatly cut back, exposing bare framing underneath. We were dumbstruck.

"Why on earth did you destroy the roofing?" I asked the flashing installer.

"I always cut back that excess roofing," said the flasher patiently, as if explaining to a Grade 2 class. "Otherwise it makes the flashing job much more difficult. Besides, the flashing is the waterproofing, right?"

This installer had been installing flashing in this fashion for years. How many roofs had he compromised?

The flashing installer was an employee. He was not fired, but he was delayed until the roofing was repaired, and I expect the roofer back charged his company.

What are the principles? What does this Tale tell us? Two things: 1). If we had good inspection records from the successful inspection (and we did) and had there been a subsequent failure, an investigation and the cut back roof came to light, our good records would have probably protected us from litigation. The developer forgetting his clipboard on the roof was dumb luck. 2). The easiest way to replicate that dumb luck is to deliberately go back to areas that have been "passed", looking for items that may have occurred since your original review.

Why field review?

To catch the contractor's after thoughts

Figure 140 - Afterthoughts cut into the Roof Later

RECORD the journey	Photograph inspected conditions	Summary Principles & Tools	
RESOLVE the issues	Use i-A's to track late issues	i-WorkPlan (i-WP)	Specific procedure including call back for late work
REVIEW the results		i-Actions (i-A)	For each discovered deficiency
REMEMBER & learn to improve		i-KnowHow (i-KH)	For each typical late issue

## 48

Reviewing Construction
Reporting & The Value of Your Time

"Okay, I get that this is your standard report, but I only need two facts and I can't find them."
– Architect to Testing Agency

What's the point? How many times have you received a competent but incomprehensible report? My favorites are concrete test reports:

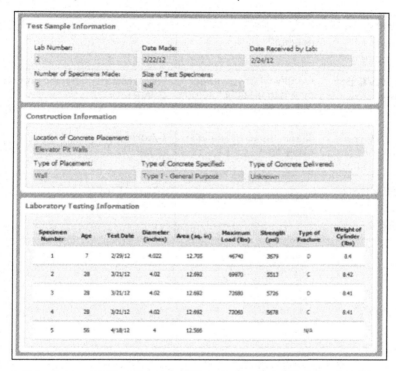

Figure 141 - Concrete Test Report - Where's the Specified Design Strength?

What's the principle? Unless you are the structural engineer, all you generally want to know is a). What was the specified strength, and b). Did the break sample meet or exceed the specified strength and/or is it likely to reach it in the specified time frame. That's it. Yet the report will be replete with esoteric data designed to protect the test agency from liability in case of problems. Sometimes, as in the sample above, key data such as the specified design strength is entirely missing!

You have to look at hundreds of these over the course of a big project. Even worse is when they are transmitted with a cover letter or email that does not cut to the chase – DID IT PASS?!

If you have to review 500 concrete test reports over a project and you spend even 1 extra minute per report finding what you need, you have just wasted $1200.00 of your fee (500 min is approximately 8 hours @ $150 = $1200).

What's the best practice? So when you receive the first of the many reports that you will need to review as managing/prime consultant, identify just what data you need. Mark that up on the report, then scan and email it back to the sender. Call them and politely ask them to either a). Give you the information you have highlighted in the cover transmittal or email, or b). highlight that information in every report they send you.

So long as you are prepared to reciprocate by occasionally highlighting content of your reports or synopsizing content in the covering email, you will all save $000's.

RECORD the journey	What records are needed?	Summary Principles & Tools
RESOLVE the issues	Identify what you need others to identify for you	i-WorkPlan Specific procedure (i-WP)
REVIEW the results	Check those essentials every time	i-Actions (i-A)
REMEMBER & learn to improve		i-KnowHow (i-KH)

## 49

Reviewing Construction
The Architect as Teacher & the Curse of Piecework

"I'm sorry, our company is only drywalling the 'B' suites."– Installer to Architect

"I'm sorry, we missed LEED Platinum by one point because the baseboard installer substituted a high VOC adhesive because it sets faster – and didn't tell anyone!"
– Project Manager to broken-hearted LEED Coordinator

What's the point? We live in a world of increasing piecework. In the days before the industrial revolution piecework had a place, perhaps. But in those days many piece-workers were exploited – they still are.

Over the past twenty years in my geographic areas of operation, piecework has invaded almost every trade involved in building construction – I know of at least a few professional practices that have also adopted this approach for the production of construction documents. A pox on all their houses!

What's the principle? I strongly believe that one of the roles of a professional is teaching. So when I reject work on a construction site, sometimes there is an accompanying teaching moment. In the case of the drywaller first quoted above, the installation crew was inadvertently destroying the air/vapor barrier around window openings by boarding right over openings, then cutting into them with a skil saw.

So I patiently explained to the crew how to achieve the same result without destroying the air/vapor barrier. Their work improved thereafter.

The next time I came to the site, I noticed the good work in the suite I had done my teaching in had not carried over to the next suite. I asked for the foreman of the "A" suite crew so I could remind him of what his crew had learned two days ago. That was when I was patiently informed that not only was it a different crew doing the "B" suites as compared to the "A" suites, it was, in fact, an entirely different company!

Pieceworkers are the least likely to be informed about key performance criteria for building construction. The second quote above refers to an office project where we had LEED Platinum in our crosshairs. We were wrapping up the tenant improvements to the project, just needed to pass the Indoor Air Quality Test prior to occupancy. We were very confident, largely because our staff LEED Coordinator had done a spectacular job throughout the construction phase. Imagine our disappointment when the test failed! Why?

The tenant improvement carpet subcontractor had pieceworked the vinyl baseboard. Low VOC adhesives set up more slowly, so to make more money by going faster, the pieceworker, who was paid by the linear foot of baseboard installed, substituted his own adhesive, which caused us to fail the IAQ test. Contractually, we only had to earn LEED Silver, so there were no recriminations except for our LEED Coordinator's broken heart.

What's the best practice? Such is the nature of construction, especially in the residential arena. I now insist on having the project superintendent with me for all field reviews, then tell him/her explicitly that it is their responsibility to pass on any learning to every crew on a project.

As for the high-VOC baseboard glue, there is no pat answer. But high-VOC compounds usually smell more strongly than their low-VOC cousins. Follow your nose!

RECORD the journey	Identify key performance criteria that will affect acceptance or testing	**Summary Principles & Tools**	
RESOLVE the issues	Broadcast issues for contractor transmission to all trades	**i-WorkPlan (i-WP)**	Procedure for superintendent accompanying
REVIEW the results	Review for conformance to identified criteria	**i-Actions (i-A)**	
REMEMBER & learn to improve		**i-KnowHow (i-KH)**	

50

Reviewing Construction
Non-Conformances & Deficiencies

"We get all the problems fixed in the end, I don't see why we need to write them all down – it just wastes time and I already have little enough of that!"– Superintendent to Quality Manager

"Your computer tracking systems don't work for me– all the important records are in my binder."– Superintendent to quality manager, shortly before superintendent quit and left project.

What's the point? The first quote above arose while I was debating with a colleague why a good quality management system includes recording issues, non-conformances, deficiencies, etc., and in particular, tracking them to resolution. His personal approach to issues of any kind was to write them down in a small book he carried with him always, then to transcribe whatever was needed into action, usually phone calls or emails, sometimes meetings. I was explaining that's not enough and further, that his transcribing was a waste of his time.

Regardless of your role, designer, builder, client, there are few if any guidelines about how to manage the day-to-day information flow on a construction site. I have seen every conceivable approach, and almost any approach will work for the individual using it – and that's the issue.

It worked for the Superintendents quoted above to carry their thoughts and observations in a small black book or binder. But it doesn't work for anyone else on the job, just for them! What happens to those black book records if he/she is sick or leaves the project or worse, leaves the company? (See second quote above - different superintendent, same project).

What's the principle? There is often confusion on a construction site between issues, non-conformances and deficiencies, so various companies track them all the same, or all different or in various mixed up ways. To be clear, in the AAGC world:

Issues are emerging aspects of construction that may well become non-conformances or deficiencies if left unattended. For a consultant, issues might include a late or absent submittal with a schedule indicating affected construction is about to start – not a problem yet, but coming soon! For a contractor, issues might include late delivery of materials crucial to maintaining schedule.

I use the action subtypes of my i-A's to identify problems:

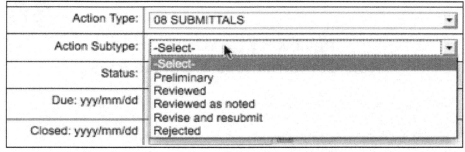

Action Type:	08 SUBMITTALS ▾
Action Subtype:	-Select- ▾
Status:	-Select- / Preliminary
Due: yyy/mm/dd	Reviewed / Reviewed as noted
Closed: yyyy/mm/dd	Revise and resubmit / Rejected

Figure 142 - Issue status by action subtype

Non-conformances are completed elements of work in progress that do not conform to the construction documents: the wrong carpet, the wrong color; the wrong air-handling unit. They are often installed well – they're just wrong! The contractor has not called for a final inspection yet, but it's clear the non-conforming elements do not conform and you don't need to wait for the end of the project to say so.

Deficiencies do not just happen at the completion of construction, which seems to be the opinion of many contractors. Anytime you are asked to review completed work, such as a floor of a building, or a portion of a floor, whatever element of the work it is that appear completed but is not correct or well done is a deficiency. If you identified the carpet as non-conforming some time ago because it's the wrong color and it was not replaced, now it's a deficiency because you were told it's complete and it's still wrong.

Observations are, by contrast, when you take it upon yourself to review an area that is still in the process of being constructed. Most things you observe are just that – observations. These are valuable and need to be included in field reviews, but they should not be identified by consultants as non-conforming or deficient, at least not yet.

What are the best practices? Information relating to building construction must reside in a location where others can see and work with it. Otherwise, what happens when you are on holiday, sick, double booked, or leave a project or company?

It's a truism that in a traditionally organized consulting practice, there is small residual value in "goodwill" and "business development" because the brains of the organization head out the door every day. What if the principal is hit by a bus? Huge amounts of business information are lost.

The same is true on a construction site. Construction can be quite volatile. It is probably as important that construction information be kept in a location where the project team can access it and work with it. In fact, this is pretty much the case with most larger builders, but only to a point.

Whereas most construction companies use databases to manage costs, RFI's, submittals, meeting minutes, etc., few have similar arrangements for the information that arises each day on a site – issues, non-conformances and deficiencies. Prior to the arrival of tablet computers, there may have been an excuse for this, but no longer. Field staff in the field can now access all of the documents that describe what is to be constructed. They can use an i-WorkPlan to remind themselves what is current and coming, rather like a traditional look ahead schedule (A look ahead schedule is simply a detailed schedule for the next 2-3 weeks). They can use i-Actions to capture the details of emergent issues as soon as they find them, assigning responsibility before moving on to the next issue. And when they identify new knowledge or necessary refinements to established practices, they can use i-KnowHow to capture that information for themselves and others.

There are lots of issues, non-conformances and deficiencies in construction – is it ever desirable to enter that data more than once? Of course not!

The i-Action below includes something I like to incorporate where possible – short form instructions in the center column. I developed this format on a large, complex project where many different individuals with many different job titles working for many different companies were all identifying non-conformances and deficiencies. Oh yes, and because of the long duration of the project the players were changing regularly.

Try as we might we could not train the dozens of folks involved in identifying and managing non-conformances. So we added an "Instructions" column that helped immensely. My assumption was that users would delete the instructions column when they had filled in what they needed to on the form, but they never did. And of course leaving the instructions never did any harm.

Item	Instructions	Record
Description of deficient condition	Summary	Incomplete rebar L2 Grid 5c
Originating Inspection Report	Reference # and issuing company	Structural 453-2
Document reference	Drawing/sketch #, spec section, etc.	S14
Location	level, grid coordinates, elevation, etc.	see above
Immediate Action(s) required (if any)	as noted by Consultant/Inspector or Contractor personnel	complete rebar and have structural engineer sign off
Other work affected by this deficiency	(See Note 1 below)	
Trade sign-off (where required, sign at right)	Indicate Company and print name	
Approval Sign-Offs:		
Contractor	Name:	Signature:
Consultant	Name:	Signature:
Owner	Name:	Signature:

Figure 143 - Non-conformance Notice (NCN) with Instructions and Sign off

In reviewing this and the subsequent deficiency i-A in the light of successful project completion, there are a number of aspects that could be and were improved as the project progressed:

At the start of the project the consultants were worried about using the contractor's Non-Conformance Notice (NCN) form and continued to use their own report formats. Hence the reference to the consultant's originating report. Where there was no originating report this part of the form was just left blank.

The "immediate work required" and "other work affected" information are primarily for the contractor and go to delays and changes to the schedule and rearranging scheduled work to accommodate problems.

Depending upon the project, not all of the sign-off's might be used for every item. It is testament to how quickly technology is changing that by the end of the project some consultants and trades trusted that just having their names printed on the form constituted adequate sign-off.

In terms of reporting requirements, there is generally no difference between a construction deficiency and a non-conformance notice (NCN) report.

In a previous Tale we dwelt on the use of a standard field review report, which allows consultants to report observations as well as non-conformances and deficiencies. The disadvantage of this approach for the contractor is that single reports must be disseminated to multiple trades, as many as are mentioned in the report. The resolution of issues across multiple trades becomes very complicated.

The standalone NCN's and Deficiency forms noted above will assist the resolution process. If consultants are unwilling to prepare standalone NCN/deficiency i-A's, and don't choose to use the same i-A tools as suggested in this e-book, it may be worth the contractor's while to transpose their report contents to individual items for tracking purposes. Where this approach was incorporated on a large project with 100+ trades and 800+ project team members, it evolved to a point where the consultants used the contractor i-A's to track and record NCN's and deficiencies, rather than their own reports. How ironic!

RECORD the journey	Capture each NCN & deficiency	Summary Principles & Tools	
RESOLVE the issues	Each i-A stays open until signed off	i-WorkPlan (i-WP)	i-A for NCN's & deficiencies
REVIEW the results	Provide place for sign-off	i-Actions (i-A)	
REMEMBER & learn to improve		i-KnowHow (i-KH)	

Links: Go to http://bit.ly/aagc50-1 for a sample non-conformance notice.

Go to http://bit.ly/aagc50-2 for a sample deficiency notice.

Note that these are fairly heavy-duty samples. Read the content carefully and edit as appropriate for your practice.

## 51

Reviewing Construction
High Volume Deficiency Tracking

"At an average of 10 deficiencies per room, 100 rooms per floor and 1 floor scheduled for completion each week, we have no tools deployed as yet to record and manage the volume." - Quality Director to Construction Manager

What's the point? On that same large project, as we approached completion, we realized that none of our pre-existing contractor systems would efficiently manage the deficiency lists associated with 6,000 separate rooms, 30 active trades, etc., even if the consultants kept up their deficiency reviews with the pace of completion, which would be a challenge in itself. Nor could we manage the even higher volume of deficiencies we would recognize and have trades resolve before the consultant team was asked to attend on site. We began to search for a solution to high volume deficiency management.

What's the principle? Clients, consultants and contractors all look at the completion of construction differently. The client wants to move in or have buyers or tenants move in, and both the client and those end occupants will want to ensure the work is to their satisfaction. Meanwhile, the consultants are balancing client expectations, their own desires for their designs and the practicalities of building code and planning requirements. And the contractor wants a completed building to be proud of and hopefully to earn a profit from.

For larger or more complex projects, everything is further complicated by the sheer numbers – of floors, of rooms, of trades, of consultants, of owners and tenants. Set against this, all deficiencies must be completed by hundreds of individual workers employed by two or three dozen main subcontractors, perhaps hundreds of sub-subcontractors and suppliers.

Traditionally, clients and consultants walk through parts of buildings after contractors indicate those portions are complete, identifying deficiencies as they go. Because it is not their responsibility to assign specific responsibility, their lists tend to be by room and floor, mixing trades and suppliers without regard to the niceties of contractual arrangements between the general contractor and the others responsible for construction. In fact, there is nothing in most consultant contracts that requires them to consider deficiencies on a trade-by-trade basis, even though this is essential to the contractor's management of the completion process.

On smaller projects, it may work to have the contractor transcribe various styles of report from a variety of consultants and owners onto a spreadsheet or similar device, then "slice and dice" the list by trade or location. But on larger projects, the constraints become unmanageable using conventional tools. Often the consultants cannot keep pace with completion using conventional note taking, the contractor cannot transcribe and distribute deficiency lists fast enough, and the individual workers do not have a ready way of communicating that they have completed their work.

It was becoming apparent that the ideal system would allow the contractor, then the consultants, then the client and occupants to successively walk through the building and quickly identify a continually decreasing number of deficiencies. Then the system would divide the deficiencies amongst the trades, preparing individualized lists, including a space for the workers in the field to sign off as each item was completed. Then the contractor, consultants and/or clients and occupants would be able to revisit items and sign them off.

To further complicate the process, based on corporate experience, the system would need to accommodate at least 50 trades and 800 or more "typical" trade deficiencies (and how sad is that!), as well as unique deficiencies based on the particular design. Managing those 50 trades and 800 deficiencies is what separates high volume deficiency tools from also ran's.

What are the best practices? After some investigation, we found such a specialized software, called Ident. Ident is part of a class of software that works this way:

First, you upload plans of your project's rooms, or floor, or elevations – whatever you want to mark up with deficiency information. In our case this amounted to 700 drawings, which our amazing summer student accomplished in two working days.

Next you upload a list of standard deficiencies. In our case, our master list of all the things that typically can go wrong – remember those 50+ trades and 800+ named deficiencies noted above.

Finally, as you walk about a building with a tablet, you call up the area you are walking through, identify a trade and deficiency type, touch the screen in the location of the issue, complete a short dialog about trade responsibility, custom notes and attaching photos, and the deficiency appears as a red dot on the plan with an associated tracking number (when an item has been confirmed as closed, it changes color/shape and can be left visible or hidden).

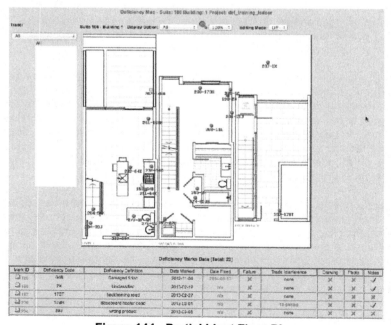

Figure 144 - Partial Ident Floor Plan

On the large project mentioned at the start of this tale, each member of the consultant team was able to record 200-300 deficiencies per day, well in excess of what they could handle using conventional handwritten lists.

What contractors love about Ident is that its default printout easily and readily separates deficiencies by trade, something consultants would generally not agree to do. Our quality management team automated the process so that they required only three hours to prepare and distribute to each of up to 50 trades an entire floor's worth of trade-specific floor plans with deficiencies noted on the plans. As part of the team effort, the architects agreed to use Ident provided that we substantially rewrote our standard deficiency list to match their preferred wording – a small price to pay for the acceleration of the process.

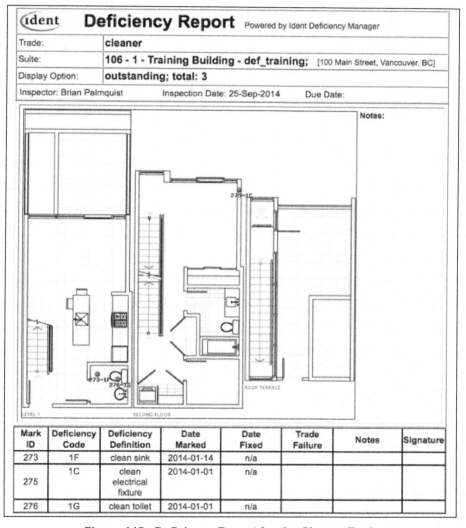

Mark ID	Deficiency Code	Deficiency Definition	Date Marked	Date Fixed	Trade Failure	Notes	Signature
273	1F	clean sink	2014-01-14	n/a			
275	1C	clean electrical fixture	2014-01-01	n/a			
276	1G	clean toilet	2014-01-01	n/a			

Figure 145 - Deficiency Report for the Cleaner Trade

The reporting feature makes it efficient for a contractor to disseminate custom deficiency lists to each individual trade. The PDF printout shown above can be electronically or hand annotated to show completion of items.

There is other software that operates similar to Ident, although I still prefer its simplicity of use, robustness and price point (Honest, I'm not getting a dime for saying this!). If you are evaluating software for non-conformances and deficiencies, you may find the evaluation checklist below to be helpful. But if a nominee won't handle all of your trades (50+) and all typical trade deficiencies (800+) in a simple, efficient manner, it's a waste of time and money – that's the "demo" to demand, all else is superficial smoke and mirrors:

				Deficiency Software Evaluation	
Yes	No	?	N/A	Item	Comments
				iWorkPlan	
				instantaneous	Web-based
				interactive	is a Database; can be printed out; can be used by many users on many projects.
				individualized	standard content can be supplemented by content customized to client, consultant, contract type, community, climate, construction type
				informative	diary space for each deficiency; open/closed status easily visible
				issue oriented	emerging issues can be associated with standard tasks
				incomplete	can be updated "on the fly" as circumstances change
				iTraveller	
				immediate	items assignable by immediate email in office and in the field
				identifiable	issued items have distinctive and informative appearance
				individual	items assignable to an individual; can be reassigned as work flow progresses; attachments collected so visible to all; assignee and assignor reminded of due date; user can see all items assigned to/by him/her;
				inclusive	items have attached forms and templates where applicable, e.g., RFI, Submittal review
				iKnowHow	
				inclusive	expertise database maintained by project users from central master
				Other	
				Cost	cost effective solution;
				Compatibility	with which tablets and smart phones? iPad? Android? Windows?;
				Training	instructions available online and logically organized;
				Storage	secure and backed up; synchronized between office and field; database independent of hardware devices
				Old school	supports form/template print out and fill in by hand
				Cross platform	Can the application import/export data from other applications? Which? Easy?

Yes means software provides functionality, No means it does not; ? means unsure; N/A = Not Applicable; Item = Item to be considered; Comments = by D, B, C or others.

Figure 146 - Deficiency Software Evaluation Checklist

RECORD the journey	Upload drawings of everything you need to review	**Summary Principles & Tools**	
RESOLVE the issues	Each deficiency report stays open until signed off	**i-WorkPlan (i-WP)**	
REVIEW the results	Trades sign off completed work	**i-Actions (i-A)**	To transmit Ident plans to trade offices

Links: To find out more about Ident, contact Richard Stratford at richard@identsoftware.com. Go to http://bit.ly/aagc51-1 for a copy of the AAGC software evaluation checklist.

52

Reviewing Construction
Why go Early to the Mockup?

"We always tilt the windows onto one corner as we take them off the dolly. How else can two guys manage them?" – Installer to Architect

What's the point? I have found that going early to a mockup is an excellent investment of time for what it teaches me. On this occasion, a window water penetration test combined with a visual mockup, I went early so I could see the window installation process – many components of window installation are invisible once completed.

The window being tested was almost 3 square meters (25 sq.ft.) – quite a heavy item when double-glazed in a metal frame. I arrived as the two-man work crew was dollying the window to the installation location – standard procedure.

Imagine my surprise when worker 1 tilted the window carefully off the dolly onto one corner, so worker 2 could slide the dolly out from under. "Aren't you worried about stress in the corner of the window frame?" I asked the pair. They looked at me as if I was a complete idiot and offered the quote at the top of this tale. Needless to say, the installation procedure was subsequently changed (a third person was added). Incidentally, the window failed its first test, coincidentally leaking at the stressed corner. I would have had no explanation for the location of that leak had I not seen the initial installation sequence.

What's the principle? In fairness to the workers, regular-sized windows are usually installed by a crew of two. My rule of thumb is that if the crew can carry the window they are a big enough crew. A third person may be required for larger or more complex windows. In a mockup, I watch the activities of the installers as carefully as the products, looking for: installation procedures that may stress the materials; inadvertent substitution of materials (*"We always use that sealant."*). Most important, I ask lots of questions, usually prefaced with *"Sorry for what may seem like a dumb question, but why…?"* (There are no dumb questions).

RECORD the journey	Come early to the mockup	**Summary Principles & Tools**	
RESOLVE the issues	Look for unusual procedures, substitutions, etc.	**i-WorkPlan (i-WP)**	
REVIEW the results	Use standard tests	**i-Actions (i-A)**	Mockup review report
REMEMBER & learn to improve	Feed experience back into the specifications	**i-KnowHow (i-KH)**	

Links: To access the form "Mockup record" or "Mockup Requirement Identification", go to http://bit.ly/aagc52-1 and select from the blue http links at the bottom left of the screen.

201

## 53

Reviewing Construction
When is a Mockup NOT?

"What do you mean it doesn't match the mockup we did 2 weeks ago. Of course it does! You were there! I was there!" – Contractor to Architect

What's the point? Many consultants and contractors are very good about doing mockups. Most are very bad about recording them. Capturing what actually happens at a mockup is crucial to the successful execution of a design.

Too often a mockup works like this. Workers different from those who will actually execute the work prepare the mockup. If they cannot assemble all of the specified materials they substitute what is conveniently accessible (read "in the back of the pickup truck"). At least one key component is unavailable so is jury-rigged. At least one of the client representative, architect, responsible consultant and one or more of the key subcontractors does not show up.

The mockup is prepared with no one present who will review it. A meeting of the review team takes place after the mockup is completed and the workers have dispersed. Various attendees take notes and photographs. The contractor and client believe the mockup has been approved. Weeks later one of the key participants rejects a large installation of the materials, claiming it differs from the mockup.

If I had not seen this play out so many times I would agree with you that this is a joke.

What's the principle? There are two types of mockups. Specified mockups are those detailed in the construction specifications. Constructability mockups are generally initiated by the contractor to explore how to construct a segment of the building when there is a concern about the ability to construct it as detailed (or not detailed). Other than making note of the type of mockup, the actual execution of a mockup should not vary.

There are also two "formats" for mockups. "Standalone" mockups are just that, separate constructions designed to stand apart from the building construction, ideally available for reference throughout the construction process. Alternatively, "First of" mockups occur where the initial construction in or on the building itself is taken to a certain (early) point, then paused while reviewers have a look and confirm whether the early results match expectations and specifications. If so, construction carries on to that accepted standard.

Figure 147 - Standalone Exterior mockup

One challenge of the "first of" mockup is that it may require early attendance by following trades. In the photo below, which is a standalone mockup, if the mockup had been
"first of", the mason would have been on site with the real crew proposed to complete
the work, rather than being there much earlier than would otherwise have been scheduled, probably with masons who would not be available to work on the project later.

Figure 148 - Corner Mockup detail - 3 Trades

Conversely, the value of a standalone mockup is earlier identification of issues. The roof portion below appears unresolved at the parapet (right side) edge – much better to find this out before all the roofing has arrived and the flashings been bent.

Figure 149 - Mockup of a complex roof interface detail

What are the best practices? There are strengths and limitations of both types of mockup. I have developed the following specification to clarify them:

<u>Mockups</u> are representative samples of construction incorporating both typical conditions such as window and wall penetrations and conditions identified as unique or novel for the Project. Mockups are required to verify compliance with requirements specified or indicated. These services do not relieve Contractor of responsibility for compliance with the Contract Document requirements. Please note:

a. Where space onsite is limited, make mockup the first area constructed representing the difficult typical & special problem conditions. Alternatively, use a parkade or offsite area.

b. Mockup indicated assemblies in advance of general constn. Include ALL components in phased fashion allowing Designer, Builder & Installer to review, refine and "sign-off." Preserve mockups for reuse during project.

c. Where a mockup incorporates a Design Change or proposed Substitute or Alternative, then the above noted services apply, including their conditions and limitations.

d. MOCKUP SUMMARY REQUIREMENTS pertaining to Building Envelope Professional (BEP) services:

    1) Builder/Contractor
        a) Provides reasonable notice of mockup scheduling
        b) Provides mockup location
        c) Uses same crew for mockup as for balance of related construction
        d) Advises of any proposed design changes or alternative or substituted products
        e) Does not use alternative means or methods "just because it's a mockup."
        f) Invites affected Registered Professionals to mockup, always including the Coordinating Registered Professional/ Architectural Registered Professional (CRP/ARP)
        g) Constructs mockup in progressive fashion so that relationships between products are clearly visible

    2) Owner
        a) Attends mockup review
        b) Signs off mockup when satisfied, indicating conformance to appearance/performance intentions

    3) CRP/ARP
        a) Attends mockup review
        b) Signs off mockup when satisfied, indicating conformance to appearance/performance/Part 3/ Part 5 intentions

    4) BEP
        a) Attends mockup review
        b) Signs off mockup when satisfied, indicating conformance to Part 5 performance intentions

NOTE: where subsequent construction differs from mockups, then those differences are treated as Design Changes and/or Alternative and Substituted Products.

Figure 150 - My General Mockup Specification

I have also developed a simple i-Action mockup report that looks like this:

| | | MOCKUP REVIEW: | |

This form is used to record a mockup event. To complete this form or reply to this email, select "Reply All", update information & select "Send". "B" = Builder, who indicates: "X" = details attached/ should be attached; "N/A" = not applicable on this occasion; "D" = Designer, who indicates "R" = Reviewed; "RN" = Reviewed as Noted; "RR" = Revise & Resubmit; "X" = Rejected. If you are using an iPhone or iPad you can select a "COMMENT/ SIGN OFFS" text box and dictate comments.

D	B	Description	Comments/ Sign off's
	x	**MOCKUP REQUIREMENT:** _by Spec _Constructability	
	x	Spec Number/ Section:	
	x	Mock Up Description in Spec:	
		Before attending mockup:	
	x	Has Mock Up Requirement been modified? "N" = No, Mock Up is to be carried out as described in the Spec. "Y" = Yes, Authorized changes to the Mock Up requirement are:	
		Consultant(s) have authorized changes to mockup requirements: Y/N	
	x	Mockup crew is same as installation crew: Y/N If "No", reschedule mockup when installation crew is available	
	x	Mockup is complete - no elements or finishes missign? Y/N If "No", reschedule mockup when mockup complete	
		MOCKUP REVIEW:	
		Mockup Date: 1st Review: 2nd Review:	
		Location: 1st Review: 2nd Review:	
		Attendees:	
		Consultant Comments:	
		Comments by Others:	
		Review Outcome: "R" = Reviewed "RN" = Reviewed as Noted "RR" =Revise and Resubmit - New date for review:	
		Attached Records (photographs, sketches, instructions, etc.):	

Figure 151 - Mockup Report i-Action

The Mockup Review includes basic but important information AND records it!:

Why are we doing this? Is the mockup required by specification, or has the builder elected to prepare a constructability mockup to explore with the design team something that may not be clear?

What have we done? If this is a specified mockup, which spec requires it and have we built it according to the construction documents? If not, what are the changes and have they been approved?

When did we do this, who was there and how did we do? I know this sounds ridic-ulously simplistic, but I have seen too many "I was supposed to be there, not her", or "if I was not invited it was not a valid mockup" arguments. Clearly record the complete attendee list.

Record who said what. An exhaustive transcription is not required, just key comments. If you are in doubt as to what a "key comment" is, record more rather than less.

Perhaps most important of all, record the outcome. Was the mockup "Reviewed" (read approved, except insurers and lawyers won't let consultants say that), "Reviewed as Noted" (no need to get back the consultants presuming the contractor fixes a list of minor issues), or "Revise and Resubmit" (rejected by any other name). If the latter, how soon can we meet again to see the fixes? It's just like reviewing shop drawings, except in three dimensions.

What's the Evidence? There is no such thing as too few well labeled photos at a mockup. I have seen many instances where a subsequent installation was rejected based on *"...that's not what was intended..."* pitted against *"...but that's what you approved..."* countered with *"...I was not looking at that part..."* or *"...the mockup was to look at the panel configuration, not the joint construction."*

If the photos issued with the mockup report say something like "Panel joint arrangement as reviewed" or "range of panel arrangements as reviewed", we would be more likely to identify the miscommunication.

Needless to say, issue a mockup report very soon after the event and read anyone else's report of the same event. Mockups are like meeting minutes in the sense that they are expected to be issued and reviewed promptly; and where there are differences of opinion, these are to be tabled right away. Designers need to understand that mockups are usually scheduled along with delivery dates, and delaying the one will hold up the other.

RECORD the journey	Record for each mockup	**Summary Principles & Tools**	
RESOLVE the issues	Capture issues until closed	**i-WorkPlan (i-WP)**	
REVIEW the results	Look for sign-off's from all affected parties	**i-Actions (i-A)**	Mockup report
REMEMBER & learn to improve	Capture results to specifications & other practice standards	**i-KnowHow (i-KH)**	Capture questionable systems & products to avoid future reuse

Links: Go to http://bit.ly/aagc53-1 for a mockup report template.

## 54

Reviewing Construction
The Real Lesson of a Mockup

"We're here to test the windows, not the walls. Concrete this thick can't possibly leak!"
– Contractor to Architect

What's the point? Many contractors as well as designers do not understand how little it takes for air and water to leak into a building, especially wind driven water.

I try to arrive early for a scheduled water test. That's where water is sprayed against a window under pressure that mimics a wind driven rainstorm. It's useful to see how the window has been installed to ensure it matches the details. In this case all appeared just fine.

Then I noticed a 10cm (4") long hairline crack in the concrete at a lower corner of the window. *"There's no point testing the window,"* I said to the contractor, *"It's going to fail."* He looked at me like I was nuts, because a 20cm (8") thick concrete wall could not possibly fail on account of a hairline crack.

What's the principle? Of course, it did leak, within 30 seconds of turning on the pressure fan. Had the final elastomeric paint coating been applied to the concrete as specified, it would probably have filled the crack and held out the water, but the paint wasn't there because the contractor did not believe that a concrete wall would leak through a hairline crack.

The conclusions of this exercise: an incomplete mockup (no elastomeric paint) may compromise a complex testing effort that affects scheduling of a significant part of the construction project.

What's the best practice? To reduce this incidence of time wasting, we have modified our i-A Mockup record to include a pre-mockup questionnaire – refer to the last two questions in the illustration below:

D	B	Description	Comments/ Sign off's
		MOCKUP REVIEW:	
		This form is used to record a mockup event. To complete this form or reply to this email, select "Reply All", update information & select "Send". "B" = Builder, who indicates: "X" = details attached/ should be attached; "N/A" = not applicable on this occasion; "D" = Designer, who indicates "R" = Reviewed; "RN" = Reviewed as Noted; "RR" = Revise & Resubmit; "X" = Rejected. if you are using an iPhone or iPad you can select a "COMMENT/ SIGN OFFS" text box and dictate comments.	
	x	MOCKUP REQUIREMENT: _by Spec _Constructability	
	x	Spec Number/ Section:	
	x	Mock Up Description in Spec:	
		Before attending mockup:	
	x	Has Mock Up Requirement been modified? "N" = No, Mock Up is to be carried out as described in the Spec. "Y" = Yes, Authorized changes to the Mock Up requirement are:	
		Consultant(s) have authorized changes to mockup requirements: Y/N	
	x	Mockup crew is same as installation crew: Y/N If "No", reschedule mockup when installation crew is available	
	x	Mockup is complete - no elements or finishes missing? Y/N If "No", reschedule mockup when mockup complete	

Figure 152 - Is the Mockup Ready & Complete?

RECORD the journey	Ask questions about readiness before going	**Summary Principles & Tools**
RESOLVE the issues	Capture issues until closed	**i-WorkPlan (i-WP)**
REVIEW the results	Look for sign-off's from all affected parties	**i-Actions (i-A)** Mockup report
REMEMBER & learn to improve	Capture results to specifications & other practice standards	**i-KnowHow (i-KH)** Capture specific mockup preparation procedures

Links: Go to http://bit.ly/aagc53-1 for a mockup report template.

## 55

Reviewing Construction
When the Rain Comes - Mold

"I'm so excited! I've identified two new species of mold never before seen!" – Industrial Hygienist to Architect

What's the point? It was one of our first building envelope remediation commissions. Frighteningly, it was for a building still under construction.

We had been called in to examine Phase Two of a wood frame project under construction, because Phase One, completed, sold and occupied, was evidencing signs of building envelope failure – mold-like substances at the base of walls and a few spots on some ceilings.

Notice that I used the term "mold-like substance." That is deliberate. Learn it. Unless you have a Ph.D. in the subject, you can identify that something looks like mold, but you cannot verify that it is mold. "Ordinary" design and construction professionals do not have liability insurance that allows them to make specific pronouncements about mold. In fact, many liability insurance policies explicitly deny coverage related to mold.

The developer was simply petrified that there would be no Phase Two sales if Phase One was not made good.

Because Phase One was all enclosed, we enlisted the help of two expert consultants to beef up our investigations. We engaged one of Canada's foremost building envelope consultants, an expert at sleuthing out the truth and finding simple fixes if simple fixes were to be found. Our second consultant was an internationally renowned mold expert from a local university, who had great expertise with moldy buildings because he had lived in one!

We first wanted to determine if we had harmful mold – there are many molds, but only a few are known to be toxic. Unfortunately there are so many unknown molds (see quote at the top) that it is often impossible to determine.

The good news is none of the likely readers of this e-book are mold experts. Provided we remember that, we will likely keep out of excessive trouble in relation to water intrusion and its marker, mold-like substance growth.

This is not to say that designers and builders can ignore water intrusion and its results. But unlike with asbestos, proving that mold growth leads to liability (read $) is rather more difficult. The reason for this is that there are hundreds of molds, most with unknown health affects.

A couple of days after my expert took his samples from the building, he called me in a state of high excitement to report he had discovered not one but two new mold species never before described. Such is the breadth of the field. Of course, in the circumstances he had no idea if they might have adverse health affects. That might take years to determine.

Conversely, asbestos is asbestos is asbestos. It has a few forms, none of which are benign if disturbed. There are proven linkages between the presence of asbestos and adverse health affects, therefore its discovery may clearly lead to litigation. By comparison, the fact that there are so many types of mold makes it very difficult to ascribe poor health outcomes to mold present in building construction – but there are legions of lawyers trying.

What are the principles? There are just a few conditions that can cause the growth of mold on or inside buildings:

Firstly, the materials may be too wet. The limits of moisture content above which mold may grow are well known for most construction materials. To prevent materials from getting wet may be as simple as keeping them off the ground plane and covering them with a waterproof cover. Unfortunately, this does not address what conditions the materials were stored/transported under prior to arriving on site. This lack of provenance explains why contractors and envelope consultants measure moisture content after the materials are on site, and also often after they are installed but not sealed in.

D.	Lumber plates:
	1. Use preservative treated sill plates, set atop continuous 1/4""polyethylene sill gasket, density 1.1 as manu. by ProPack, Delta (946-9116). If concrete is uneven, use multiple layers of gasket to achieve air seal. Seal inside wood-to-concrete junction continuously with Chem Calk 900 or other preapproved sealant.
	2. Where using a pressure treated (preservative treated) sill plate, the moisture content (m.c.) of the plate may be as high as 25% so long as the M.C. of ALL abutting wood, including sheathing, is as much below 20% as the plate is above 20%. For example, a 25% preservative treated plate is acceptable where the balance of adjoining lumber is ≤15%.
	3. To protect sill plates from taking up water from concrete topping, install a continuous piece of plate polyethylene (typ. 16" wide) to cover the inside face of the plates & the partial perimeter of the floor sheathing just before topping is poured.
	4. To keep wood-frame walls drier and facilitate air/vapour barrier continuity, drape the top plate of each wall with plate polyethylene.

Figure 153 - Specifications can Allow for Wetness - see Item 2

Secondly, materials may get wet after installation. This can occur if they are not protected after installation, or if the detailing of the building directs water onto them. The first condition is the contractor's responsibility, the second is generally the designer's. However, the designer in a wet climate can anticipate that materials will get wet and specify accordingly:

2.4	TREATED MATERIALS
A.	Preservative-Treated Materials:
	1. Outside the plane of the moisture barrier, wood noted as preservative treated (preservative treated) to be PWF or K-33, CCA or ACQ treated to CSA 080, 6.4 kg/cu.m., by Western Cleanwood Preservers Ltd. or app'd alternate, kiln dried AFTER treatment including but not limited to:
	a. Strapping for drainage cavities to be preservative treated plywood to PWF standard
	2. Inside the plane of the moisture barrier, wood noted as preservative treated (preservative treated) to be borate pressure treated wood 2.7 kg/cu.m. B2O3 , Advance Guard by Western Cleanwood Preservers Ltd. or app'd alternate, kiln dried AFTER treatment including but not limited to:
	a. Wood members in contact with concrete, including exterior wall floor plates and all interior load bearing walls;
	b. Concealed members in contact with masonry or concrete.
	c. exterior columns, balcony divider walls, etc..
	3. Boracol 20 for mold touch up:Where plywood, OSB or dimension lumber is inadvertently installed with a small amount of surface mold, spray the area affected with Boracol 20 preservative & allow to dry before covering.

Figure 154 - Materials Specified Based on Exposure

Thirdly, leakage can occur after installation, such as failure of a plumbing pipe under pressure test. For that possibility, refer to the checklists below.

The best approach will always be to prevent water intrusion and where it occurs, dry out the building as quickly as possible and seek expert opinion.

Inspection timing:

- When weather reports indicate approaching inclement weather:
- At start of deliveries of each type of materials:
- At start of work of each trade:

Work in progress on site:

Storage of building materials before installation

- off ground:
- under cover or tarped:
- dry storage locations for porous building materials such as drywall:

Landscape storm water drainage away from buildings

- temporary, during construction before final gutterage installations:
- permanent installations:

Site operations in progress:

Control water use/spillage from water generating processes

- concrete and mortar mixing:
- water testing of windows, etc.:
- tile and stone cutting:
- concrete cutting:
- painting:
- plumbing testing:

Before leaving site each day

- water sources such as faucets turned off:

Figure 155 - Mold Prevention Checklist i-A – Top

What are the best practices? As you might guess, the complexity of mold has created a relatively complex checklist environment to prevent, identify and manage it. The i-A above, the top half of my full Mold Prevention i-A, addresses weather events, material storage, site drainage and manmade water incursion arising during testing.

The bottom part of the Mold Prevention i-A is below and addresses the balance of typical design-caused and construction-related sources of water intrusion. It's a fairly exhaustive list. Your project may be simpler such that you can confidently eliminate some of these items, but perhaps better to err on the side of caution.

Work in Progress building Inspection - general:

Review locations of potential water sources

- plumbing and fire suppression systems for obvious leaks:

Proper ventilation/exhaust from moisture generating appliances, esp. clothes dryers.

- vents unobstructed and straight runs:
- vent covers removable for future lint removal:

Drains, gutters, downspouts etc. operational and debris free

- temporary diverters during construction:

Roofing tie in's

- diverter flashings to take water away from walls:

Wall and roof opening protection

- roof penetration flashing, temporary cover during construction:
- wall penetrations, including doors, windows, vents, outlets, hose bibs, etc.:

Building envelope:

- caulking – adhesion, cracking, settlement:
- doors and windows – complete unit installation end of each day, or temporary cover (no unprotected openings):
- Cladding and trim – continuity and transitions complete or protected:
- Flashing installation and continuity:
- Evidence of water intrusion – report immediately:
- Evidence of finished floor or ceiling damage – report immediately:
- Pooling of water on a flat roof – more than ¼" or more than 1 day:
- Other evidence of improper drainage:

Evidence of incomplete or absent barriers – if you don't know what these are, ask!

- Air barrier:
- Moisture barrier:
- Vapor barrier:
- Thermal – insulation properly installed where required:

Figure 156 - Mold Prevention Checklist i-A – Bottom

A few years ago I was concerned about the take up of water in a wood frame arising from the widespread use of concrete topping, which is the layer of 1-1/2" (38mm) +/- lightweight concrete often used to improve acoustics and provide a better base for installation of brittle materials like floor tile. So I measured the water content throughout the bottom floor plates of one floor of a multifamily residential building under construction, just before the topping was due to be poured. The good news was the moisture content was 12-18% (anything below 19% is considered reasonable; above 25% we say mold spores are about to dine on the wood frame).

The bad news emerged when I came back two days after the topping had been poured. The bottom plates I was worried about (because they had no exposure to air once they were encased in topping) had increased in moisture content by 5-15%, measured against their pre-pour moisture content. Most of the wood was too wet to enclose with drywall.

Because I had predicted this outcome, I had worked with the builder in advance to select an area whose schedule did not require rapid enclosure. We monitored the conditions and did not proceed until all was dry enough.

The solution to this problem was simple and cheap. As we prepared for the topping pour, we stapled a short width of polyethylene (called "plate poly") to the edges of the exterior bottom plates that were going to be otherwise wetted, running the plate poly about ten inches (25cm) onto the floor and stapling it in place. We measured water take-up after topping pour on wood protected by plate poly – it was negligible.

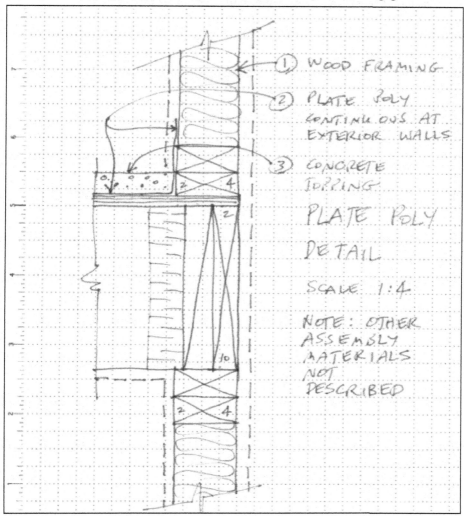

Figure 157 - Plate poly to protect bottom plates

If you are wondering why the plate poly is only attached at exterior walls, that is because in my northwest coast climate, exterior walls often include a polyethylene vapor barrier that prevents wet wood from drying, including bottom plates. Interior walls do not generally have polyethylene vapor barriers and are warmed, so their wood will dry out before mold can form.

Needless to say, the balance of the project, as well as my future projects, proceeded with this cheap but effective detail.

3.	To protect sill plates from taking up water from concrete topping, install a continuous piece of plate polyethylene (typ. 16"" wide) to cover the inside face of the plates & the partial perimeter of the floor sheathing just before topping is poured.

Figure 158 - Specification Around Plate Poly with Concrete Topping

Since water may infiltrate in many situations, checklists may remind the reviewer where to check for the most likely areas of elevated moisture content:

Weather: ☐ Sunny ☐ Cloudy ☐ Mixed ☐ Rain showers ☐ Rain ☐ Snow ☐ Other/details:_____

Day before:	Weather:	Temp.	degC High	degC Low
To-day:	Weather:	Temp.	degC High	degC Low
Day after:	Weather:	Temp.	degC High	degC Low

Ref. Dwgs:		Grid Refs:	
Report date yymmdd:	Review date yymmdd:	Associated Quality Report #:	

Area[s] reviewed/MC observed:

Floor sill plates - __%

Floor plywood sheathing (where not covered by topping) - __%

Sheathing @ inside near floor line - __%

Sheathing at outside - __%

Sheathing at centre ply - __%

Window/opening sills - __%

Window/opening headers - __%

Inside ply - __% _Outside ply - __%

Holes drilled in header? Yes __ No __ - Where?

Outside sheathing - __% _Outside ply - __%

Other locations: - __% Okay to cover? Yes___ No___

Figure 159 - Moisture Content Checklist

Notice at the top of this checklist the weather for the day before review, and the forecast for the day after, are to be logged. In addition to raising a possible red flag around poor weather, these records serve to gather some rudimentary data around the moisture readings.

Whether you are a designer or builder, if you are serious about managing the moisture in construction, invest in a serious moisture meter:

Figure 160 - My favorite analog moisture meter

I like this analog meter for two reasons: 1). It is cheap, $265 versus $400+ for a digital meter; and 2). I am more in control of what it says. Periodically I visit a construction site that has bad moisture management strategies, but am expected to measure moisture content nonetheless. 19% is the magic number for wood materials in Canada – at or below that limit you are "good to go."

With a digital meter, when you insert the probes into wood, the meter pauses a moment, then gives you a readout in huge numbers that are impossible to hide:

Figure 161 - Digital moisture meter readout

With an analog meter, the needle swings to the moisture content reading as soon as you insert the probes, and swings back to zero as soon as you remove the probes. This makes for quicker measurements when you are doing many of them because the meter does not pause a few seconds before resetting itself. That may sound corny, but

217

if you are doing hundreds of measurements in a day, as I have done, the seconds add up. Also, if you are unhappy about contractor moisture management strategies, you can decide marginal results are not good enough because the reading is gone before others can see it – devious but occasionally useful.

By the way, it is not necessary to record the location of each and every measurement. The Moisture Content Checklist (Figure 159 above) reminds you where to check. I frequently identify a location such as a room or suite, then fill in the checklist with comments like, for example for "Floor sill plates", "Suite 101, 10 dispersed locations, all below 19%", because that is the key information, all below 19%. However, if there is a problem area, I would add: "NW corner of master bedroom, 1 reading 23% - further drying required." I try to make it a habit to have the project superintendent with me when I make such measurements, so that she/he can direct immediate supplementary drying such as portable heaters in any specific areas that are too moist.

In the sad event that your design and/or the construction fail to keep water out, it is essential that the builder be proactive in scoping out the problem and quickly remediating it. "Time is of the essence" was never truer than in cases of water intrusion. Most remediation measures escalate (i.e., become substantially more expensive) after 24 hours.

The water intrusion investigation report below is in two parts because of its length. That in itself should be sufficient incentive to act preventively and proactively.

I have not magically created this report. It comes from my own research into the subject and draws from several insurance company and legal documents (I have a three inch binder on the subject in my bookcase). It is as concise as I can make it.

Water Intrusion Incident and Investigation Report		
Item	**Notes**	**Instructions**
Time		Indicate time and date when the item was first noticed.
Site Location		Indicate here the address of the building, remembering that in multi-building projects there will be more than one address. Use the address that will be easiest to find right now, e.g., "3800 Balfour Ave., Block B."
Nature of Observation - indicate which one(s) apply and provide details	• Leakage from roof: • Drain overflow: • Plumbing leak: • Window or openings in façade: • Pooled water: • Condensation from pipes or HVAC: • Mold observed on building materials: • Flooding from surrounding area: • Other (Please specify):	Indicate which, add new categories if necessary. If more than one, indicate all
Location of observed water/ moisture/ mold		Be specific. Indicate where on the site the problem was observed - be as specific as possible, i.e., which floor, which wall, which grid reference?
Suspected cause(s)		

Figure 162 - Water Intrusion Incident Report i-A – Top

The questions in this report are all straightforward but are exactly the things that are frequently not documented, which can result in very unhappy results later.

The bottom half of the form (below) documents typical specialist consultant interventions. They are rules of thumb borne out by the experience of the New York City Health Department, which was the first public body to focus on this issue:[1]

If it (the mold-like substance) is less than 10 square feet (1 sq.m.), don a mask and gloves and wash it with bleach; if larger than a table but smaller than my office (10-100 sq.ft., 1-10 sq.m.), proceed as above but use dust suppression methods and tools; if more that 100 sq.ft. (10 sq.m.), bring in the experts in their Tyvek suits and respirators, who will perform full remediation. Where mold growth is discovered in HVAC systems, the threshold areas are 1/10th those noted above (mold in HVAC systems is much more serious).

Water Intrusion Incident and Investigation Report		Upon receipt of the reported water intrusion occurrence or an incident report, the Project Superintendent should investigate the incident, document the findings and corrective actions and file the following report:
DESCRIPTION OF PROBLEM		Note condition observed, specific location, and area affected (quantity)
ACTION REQUIRED		
DATE ACTION COMPLETED		insert date
FURTHER ACTION OR EVALUATION NEEDED		Indicate if further action is required. Do not mark this Action as 'closed' until there is no need for further action
ENVIRONMENTAL CONSULTANT REQUIRED?	_YES _NO	Indicate reasons
CONSULTANT CONTACTED		Insert date of contact, including attempted contacts, e.g., "left urgent voice mail 081128"
CONSULTANT VISIT DATE		insert date(s)
CONSULTANT OBSERVATIONS		Copy here or attach separate report to this Action, including any photographs
DRYING/RESTORATION/ REMEDIATION CONTRACTOR SERVICES REQUIRED?	_YES _NO	Indicate reasons
CONTRACTOR CONTACT		Provide full contact details
CONTACT DATE		Insert date
SUMMARY OF ACTION TAKEN		Copy here or attach separate report to this Action, including any photographs
WAS NEW KNOWLEDGE LEARNED ABOUT WATER INTRUSION, MOLD OR THEIR REMEDIATION?	_YES _NO	If yes, convert this Action Item into a Knowledge Item and add to the company knowledge database.
OTHER OBSERVATIONS:		

Note: Use the top part of this form to report observations of water intrusion/excessive moisture/visible mold within the building. Use the bottom part of this form to record the investigation and resolution of the matter. This report may be completed by any employee of Contractor, or their sub-contractors. A completed report form should be submitted to the Contractor field office, either on paper or electronically. The person initiating this report is the person noted below as "Referred by" The responsible Project Superintendent should be designated as the person "Now responsible". Either fill in the form electronically, or make a printable version and fill in by hand.

Figure 163 - Water Intrusion Incident Report i-A – Bottom

By the way, in the case of the tale of leakage and indeterminate mold mentioned at the start of this tale, we were fortunate. For the building already completed, our team used a checklist for residents to help identify potential problem areas, then investigated all their reports, identifying which were actual building envelope issues. We also did our

---

1   New York City Department of Health & Mental Hygiene, Bureau of Environmental & Occupational Disease Epidemiology, "Guidelines on Assessment and Remediation of Fungi in Indoor Environments," January 2002, updated November 2008 at http://www.nyc.gov/html/doh/html/environmental/moldrpt1.shtml

review of the exterior. From this information we identified a relatively small number of issues that the developer remediated at no cost to the owners beyond inconvenience. For the Phase Two building under construction, we reviewed the design and made a few detailing suggestions; worked with the superintendent to improve housekeeping such as keeping materials dry; provided moisture content and other reviews as the building was constructed; tested the windows and the suites themselves for air and water leakage; and provided a building envelope maintenance manual for the building. This became the first of many projects with that developer.

RECORD the journey	Many moisture readings	Summary Principles & Tools	
RESOLVE the issues	Good housekeeping	i-WorkPlan (i-WP)	
REVIEW the results	Report for each area reviewed, sign-off for any areas remediated	i-Actions (i-A)	Checklists & intrusion reports
REMEMBER & learn to improve	Modify i-WP's and 10C's in accordance with experience with contractors & locations	i-KnowHow (i-KH)	Capture problem situations for future prevention

Links: Go to http://bit.ly/aagc55-1 for a building envelope condition assessment report designed to be completed by building occupants.

Go to http://bit.ly/aagc55-2 for a Mold + water intrusion prevention checklist.

Go to http://bit.ly/aagc55-3 for a Water Intrusion investigation report template.

## 56

Checking
A Checklist for Checklists

"What's the point of these checklists anyway? They are way too long and we already do the stuff on them – this is just a waste of time!" – Junior superintendent to quality manager

The concrete pre-pour checklist in question was not one of mine, but could have been until recently. A separate checklist was required for every concrete pour, however small. It required multiple sign off signatures and ran to several pages – being a general checklist it needed to accommodate every possible combination of work associated with concrete. Some was for review by the concrete supplier, some by the rebar placing foreman, some by the formwork foreman, some by the superintendent – impossible to expect a clean, consistent result.

We eventually improved things somewhat by splitting concrete pre-pour into concrete formwork, concrete rebar, concrete mix design, concrete placing, concrete slab on grade, concrete suspended slab and concrete columns and walls. Each checklist is shorter but more are required – still hard to manage. However, as the type of periodic catastrophic building failures that occurred in Tale #1 (none of mine!) underline, the consequences of a lack of diligence can be significant. No one has yet developed a construction quality management system that does not involve checklists, however that does not mean that checklists should not be designed, just as the buildings they help construct are designed.

What's the point? Many checklists are used in design and construction. Most are badly done, including (until recently) some of those authored by me. I was getting increasing suggestions and complaints about checklists designed or managed by me and used in our work. So I did some research and happened across a great book called "The Checklist Manifesto" by neurosurgeon Atul Gawande[1]. This gentleman dramatically reduced accidents in a variety of hospital settings starting with operating rooms. He did so by researching the design and implementation of checklists, paying attention to the research and applying it to his workplace, the hospital.

Checklists designed after Gawande's approach are shorter, simpler, progressive and logical – and they are more likely to get used more frequently.

Here is Gawande's checklist for checklist design:

---

1   Gawande, Atul, "The Checklist Manifesto: How to get Things Right," Picador, 2010, New York

01 Development

☐ Do you have clear, concise objectives for the checklist?

IS EACH ITEM:

☐ A critical step and in great danger of being missed?
☐ Not adequately checked by other mechanisms?
☐ Actionable, with a specific response for each item?
☐ Designed to be read aloud as a verbal check?
☐ One that can be affected by use of a checklist?

HAVE YOU CONSIDERED:

☐ Adding items that will improve communications among team members?
☐ Involving all members of the team in the checklist creation process?

02 Drafting

DOES THE CHECKLIST:

☐ Use natural breaks in workflow (pause/ hold points)?
☐ Use simple sentence structure & basic language?
☐ Have a title that reflects its objectives?
☐ Have a simple, uncluttered & logical format
☐ Fit on one page?
☐ Minimize the use of color?

IS THE FONT:

- [ ] Sans serif?
- [ ] Upper and lower case text?
- [ ] Large enough to be read easily?
- [ ] Dark on a light background?
- [ ] Are there fewer than 10 items per pause point?
- [ ] Is the date of creation or revision clearly marked?

HAVE YOU CONSIDERED:

- [ ] Adding items that will improve communications among team members?
- [ ] Involving all members of the team in the checklist creation process?

03 Validation

HAVE YOU:

- [ ] Trialed the checklist with front line users?
- [ ] Modified the checklist in response to repeated trials?

DOES THE CHECKLIST:

- [ ] Fit the flow of the work?
- [ ] Detect errors when they can still be corrected?

CONCLUSIONS:

- [ ] Can the checklist be completed in a reasonably brief period of time?
- [ ] Have you made plans for future review and revision of the checklist?

Note that the checklist for checklist design follows Gawande's own recommendations about length and pause points.

Implementing Gawande's approach led me to the checklist approach described in more detail in the next tale.

RECORD the journey	Checklists for many design & construction activities	Summary Principles & Tools	
RESOLVE the issues	Identify issues and non-conformances	i-WorkPlan (i-WP)	Specific checklist procedure
REVIEW the results	Review checklists as executed for effectiveness	i-Actions (i-A)	Each checklist is an i-A
REMEMBER & learn to improve	Refine checklists based on effectiveness of use	i-KnowHow (i-KH)	Capture problem situations for future prevention

Links: You can buy Gawande's e-book from Amazon at http://bit.ly/aagc56-1

## 57

Checking:
Checklists for one and All

"Why is it designers check everything differently than contractors? It's the same build-ing?" – Junior Superintendent to Quality Director

Although I am the author/manager of more than 150 quality checklists for the construc-tion company I work with, I had no answer to this disarmingly simple question. The comment got me to thinking - will we ever communicate well if we are always using different lists to review the quality of construction?

I began to take a second look at what we do, also in the context of Atul Gawande's fresh view of the checklist landscape.[1]

Could a designer checklist that looks something like this:

8.3 Special Doors (Panel Folding Doors, Sectional Metal Overhear Doors)

before fabrication begins, submission of:
·   shop drawings
·   wiring diagrams for power-operated doors
·   folding walls:
    ·   samples of partition finish
    ·   sound transmission test data

installation:
·   entire assembly installed to manufacturer's instructions
·   electrical connections completed for power-operated doors
·   operable parts adjusted:
    ·   sound seals to acoustically rated doors
·   exterior doors:
    ·   weatherstripping forms weathertight seal
    ·   accessories, as specified
    ·   operating components lubricated/adjusted
·   field touch-up to all damaged surfaces following installation

final acceptance:
·   withheld until operation/maintenance manuals provided

clean-up:
·   all rubbish/surplus material removed from site promptly

Figure 165 – Partial Designer Quality Checklist

---

1   Gawande, Atul, "The Checklist Manifesto: How to get Things Right," Picador, 2010, New York

Be integrated with a contractor checklist that looks like this?

Done ☒	N/A ☒	DESCRIPTION	COMMENTS / SIGN OFFS:
☐	☐	Correct materials being used?	Check containers periodically
☐	☐	Use anchorage devices to securely fasten assembly to wall constructions and building framing without distortion or stress.	
☐	☐	Fit and align assembly including hardware; level and plumb to provide smooth operation.	
☐	☐	Coordinate installation of electrical service. Complete wiring from disconnect to unit components.	
☐	☐	Touch-up paint on frame and other painted surfaces in accord with painting section.	
☐	☐	Upon completion of installation, including work by other trades, lubricate, test and adjust doors to operate in accordance with manufacturer's product data. Final adjustments shall be made by manufacturer's authorized representative.	
☐	☐	Adjust door and operating assemblies	

Figure 166 - Partial Contractor Quality Checklist

I decided after some careful thought that the short and exciting answer could be "Yes!"

The contractor checklist includes preparation that the designer has no involvement in. The two lists converge around successful management of submittals and preparation of mockups. They diverge again around substrate preparation, which is the contractor's, then converge somewhat around the details of installation. And they remain convergent around completion/closeout considerations.

Several other considerations came into play as part of the design of integrated checklists. Firstly, there needed to be clear Yes/No success/failure indications. Secondly, our lawyers advised that every checklist line needed to be considered, and where an item was not applicable on a given day, it needed to be explicitly marked as such, else we might be accused at a later date of not considering something on the list. Thirdly, there needed to be a place for comments and, where requested, sign off's that the work considered by the checklist was acceptable at the time. Fourthly, the entire checklist event needed to respect the three toolsets, i-WP, i-A and i-KH. Finally, getting back to Gawande's "Checklist Manifesto", the checklist needed to be broken into manageable pieces based on the phase of work.

Out of these considerations I have started to develop a joint checklist format that includes the requirements of both designer and builder, together in a logical sequence. My hope is that designers and builders can see what is important to each other, respect that, and perhaps collaborate on field review.

Considering Gawande's suggestion that longer checklists be broken into phase-based segments resulted in the following basic four phase structure:

D	B	Done Y/N	N/A x	DESCRIPTION	COMMENTS / SIGN OFFS:
			1	Before construction of this trade/area starts:	
			2	During construction:	
			3	Before closeout in this area:	
			4	Before closeout of this trade:	

Figure 167 - Quality Checklist Four Phase Structure

In the interests of simplicity and i-A functionality, simple, standard instructions based on i-A (and for me QW) functionality were developed:

Figure 168 - Quality Checklist Instructions

Typical standard checklist content was developed for each phase of the template:

D	B	Done Y/N	N/A x	DESCRIPTION	COMMENTS / SIGN OFFS:
			1	**Before construction of this trade/area starts:**	
	x			A SQP – Subcontractor Quality Plan(s) submitted & accepted?	Open an iT for each SQP
x	x			Mockup requirements identified & scheduled?	Includes construction mockups that may not have been specified, based on designer or builder experience. Open an iT for each mockup
x	x			Submittals: Scheduled? Submitted? Review completed successfully?	Open an iT for each submittal
	x			Installation details reviewed?	What are the specified tolerances? Open an iT for each RFI
	x			Substrate construction accepted? Plumb? Level? True to specified tolerances?	
	x			Substrate conditions accepted as clean & ready for installation?	
x	x			Special inspection or review requirements identified & scheduled?	
x	x			Preceding work reviewed for completeness & readiness to receive?	
	x			Weather checked to ensure work is within specified environmental limits?	What are the limits?

Figure 169 - QC Instructions - Before Start

Notice at the top the two columns labeled "D" = Designer and "B" = Builder. Thus responsibility for review is assigned. Note that the designer's responsibilities are generally lesser, and are always matched with contractor involvement.

Remember this is a template. In each phase there may be instructions particular to a trade added at the bottom. In some checklists some of the standard elements may not be needed, or may need editing.

			2	**During construction:**	
x	x			All areas to be reviewed ready for review?	
x	x			Areas proposed for review protected from weather?	
x	x			Deficiencies from previous reviews corrected?	iT's should be closed
	x			Contractor personnel available to accompany reviewers?	
	x			Any required primers installed but not too old?	
	x			Correct materials being used?	Check containers periodically

Figure 170 - QC Instructions - During Construction

		3	Before closeout in this area:	
x			Adequate protection installed for completed work?	
x			Adjacent surface(s) cleaned?	
x			Rubbish & debris removed & appropriately disposed of?	

Figure 171 - QC Instructions - Area Closeout

		4	Before closeout of this trade:	
	x		All non-conformances, deficiencies & other issues of this trade resolved?	All iT's should be closed
	x		Any weather related deficiencies identified together with a schedule for their completion?	Open an iT for each deficiency
x	x		Deficiency holdbacks for this trade identified & agreed to?	
x	x		Closeout submittals submitted, reviewed & accepted?	Open an iT for each closeout submittal

Figure 172 - QC Instructions - Trade Closeout

Several formats can be created for specific electronic devices. For example, when translated into the 500 pixel wide i-A format preferred by some smartphones, the (partial) document below has a distinctive appearance:

TRADE :

Weather:

Ref. Dwgs: | Grid Refs:

Report date yymmdd: | Review date yymmdd: | Associated Quality Report #:

To complete this checklist or reply to this email, select "Reply All", update information & select "Send"; if you are using a dictation-equipped smartphone or tablet you can select a "COMMENT/ SIGN OFFS" text box and dictate comments.
Complete checklist progressively or at frequency established by Designer/Builder. "D" – by Designer; "B" – by Builder; "N/A" - not applicable on this occasion. "Y" = Yes, completed/accepted; "N" = not completed/accepted

D	B	Done Y/N	N/A X	DESCRIPTION	COMMENTS / SIGN OFFS:
				Before construction of this trade/area starts:	
	x			A SQP – Subcontractor Quality Plan(s) submitted & accepted?	Open an iT for each SQP
x	x			Mockup requirements identified & scheduled?	Includes construction mockups that may not have been specified, based on designer or builder experience. Open an iT for each mockup
x	x			Submittals: Scheduled? Submitted? Review completed successfully?	Open an iT for each submittal
	x			Installation details reviewed?	What are the specified tolerances? Open an iT for each RFI

Figure 173 - on Laptop - QC i-A

And the distinctive appearance as well as HTML functionality remains on tablet and smartphone copies. This is important because emerging research into the effective use of multiple hardware (laptop/ tablet/ smartphone) is identifying the importance of similar appearance and functionality regardless of device. Users are prepared to sacrifice some functionality on a smartphone's reduced screen "real estate", but they expect this to be made up with sequential screen data entry.

All Inboxes (23)

00 Quality Checklist Template - |yTunJOzYRptz01FqehSQ7g==|
September 24, 2014 at 11:01 PM

Project: copyright 2014 - - Practice Design + Construction Smarter not Harder

TRADE :

Weather:	
Ref. Dwgs:	Grid Refs:
Report date yymmdd:	Review date yymmdd:	Associated Quality Report #:

To complete this checklist or reply to this email, select "Reply All", update information & select "Send"; if you are using a dictation-equipped smartphone or tablet you can select a "COMMENT/ SIGN OFFS" text box and dictate comments.
Complete checklist progressively or at frequency established by Designer/Builder. "D" – by Designer; "B" – by Builder; "N/A" - not applicable on this occasion. "Y" = Yes, completed/accepted; "N" = not completed/accepted

D	B	Done Y/N	N/A X	DESCRIPTION	COMMENTS / SIGN OFFS:
				Before construction of this trade/area starts:	
	x			A SQP – Subcontractor Quality Plan(s) submitted & accepted?	Open an iT for each SQP
x	x			Mockup requirements identified & scheduled?	Includes construction mockups that may not have been specified, based on designer or builder experience. Open an iT for each mockup
x	x			Submittals: Scheduled? Submitted? Review completed successfully?	Open an iT for each submittal

Figure 174 - QC i-A on iPad

RECORD the journey	Similar functionality on all hardware devices	Summary Principles & Tools	
RESOLVE the issues	Standard format for designers & builders	i-WorkPlan (i-WP)	Specific procedures for checklist creation
REVIEW the results	Non-conformance & deficiency incidence should reduce	i-Actions (i-A)	i-A formats for laptop/ tablet/ smartphone i-KnowHow (i-KH) - Knowledge items to look for
REMEMBER & learn to improve	Use Gawande's rules	i-KnowHow (i-KH)	

## 58

Checking:
Designer Pre-inspection Checklist

"So I drove all the way out here because you told me the 2nd floor was ready for pre-drywall review, but it's actually just 3 of 12 suites that are ready!" – Exasperated architect to superintendent

This was not the first time this had happened on this and some other projects. And I knew that the client would be unsympathetic to a claim for additional services – he would just say, "Well, you would have to inspect the entire 2nd floor eventually." How to prevent this from happening again?

What's the point? I have written in other tales about contract provisions that help protect designers from being used by builders as quality control inspectors, inspecting each day or two's progress, which is never their intended role. But as a practical matter, I always recommend the following short over-the-phone questionnaire when called by a builder for attendance on site:

				Complete checklist progressively or at frequency established by Designer/Builder. "D" – by Designer; "B" – by Builder; "N/A" - not applicable on this occasion. "Y" = Yes, completed/accepted; "N" = not completed/accepted	
D	**B**	**Done Y/N**	**N/A** x	**DESCRIPTION**	**COMMENTS / SIGN OFFS:**
			2	During construction:	
x	x			All areas to be reviewed ready for review?	Supt to review before calling
x	x			Areas proposed for review protected from weather?	Where required
x	x			Deficiencies from previous reviews corrected?	IT's should be closed
	x			Contractor personnel available to accompany reviewers?	
	x			Correct materials being used?	Check containers periodically
x				Appears ready for review?	

Figure 175 - Designer pre-inspection checklist

The instructions and checklist items are self-evident – they have saved me and my clients thousands of dollars in otherwise wasted time.

What are the principles? When I was a designer, my standard client contract included the following clause: *"Our proposed fees are fixed for the items we have some control over (such as design and design review), and hourly for those outside of our control (such as investigation, and dealing with changes proposed by others, and tender & construction phase services). Since most clients need a budget for hourly services, we generally include such a budget in our fee breakdown."*

The purpose of this clause is to remind the client that you charge hourly for whatever you do not control. **This includes abuse of your time by a builder using you as their quality control program.** You can usually discern a pattern of time abuse fairly early, such as when you are asked to review a significant work scope, only to discover when you arrive on site that the completed scope for review is actually very limited. If this is a pattern rather than a "one of", the best approach is to include a cover note with your next invoice to the client, saying something like *"Just so you are aware, at current rates we will bill our remaining construction phase budget in [x] months as a result of inefficient requests for our attendance on site made by the builder. Thereafter our additional services will be billed hourly."*

I have used this alert a handful of times in my career. Each time the client has spoken with the builder and a more rational approach has prevailed.

What are the best practices? My standard Division 01 specification (which other architects have happily adopted) also includes clause 4 below, to ensure the principle of efficient use of Consultant time is embedded in the contract:

B.   FIELD REVIEW

1.   The Owner and the Consultant shall have access to the Work. If part of the Work is in preparation at locations other than the Place of the Work, access shall be given to such work whenever it is in progress.
2.   Give timely notice requesting field review if Work is designated for special tests, field reviews or approvals by Consultant instructions, or the law of the Place of the Work.
3.   If the Contractor covers or permits to be covered Work that has been designated for special tests, field reviews or approvals before such is made, uncover such Work, have the field reviews or tests satisfactorily completed and make good such Work at no additional cost to the Contract.
4.   **Where the Contractor requests a field review, either in anticipation of covering over work or otherwise proceeding with the balance of the Work, and the Work that is the subject of the field review is not ready for review at the time agreed between the Consultant and the Contractor, then the Owner is entitled to recover from the Contractor costs incurred by the Consultant and charged to the Owner.**
5.   The Consultant may order any part of the Work to be examined if the Work is suspected to be not in accordance with the Contract Documents. If, upon examination such work is found not in accordance with the Contract Documents, correct such work and pay the cost of examination and correction. If such Work is found in accordance with the Contract Documents, the Owner shall pay the cost of examination and replacement.

Figure 176 - Field Review Specification

In fairness to builders, the converse occurs, when the builder expends major effort on account of incomplete, unclear or missing details, or receives repeated requests to price multiple options, etc. Construction contracts usually limit the contractor's ability to recover such unreasonable costs except through a formal "claim for delay" or similar. Designers and owners who take advantage of their builder will generally pay for it elsewhere in the form of higher quotations, claims for delay, etc.

RECORD the journey	Ask before visiting	**Summary Principles & Tools**	
RESOLVE the issues	Use the pre-inspection checklist to identify what needs to happen first	**i-WorkPlan (i-WP)**	
REVIEW the results	Identify recurrent patterns of abuse & advise the client	**i-Actions (i-A)**	Pre-inspection checklist at the head of each field inspection form
REMEMBER & learn to improve	Ensure contracts & specifications reflect the players	**i-KnowHow (i-KH)**	

59

Checking:
Checklist as Quality Teacher

"What do you mean, you want me to fill out this curtain wall checklist for every room? That's nuts!" – Installer to Superintendent

On this project, the curtain wall subcontractor kept coming in on weekends to do work, and the contractor seemed unable or unwilling to stop him. My issue was the inability to review the work before it was covered up.

I took my concerns to the contractor, whose solution was to require that the subcontractor complete and sign a detailed curtain wall checklist for each and every room – not for each dwelling unit, but for each room. And not the electronic version described above, rather a paper printout version. This amounted to about 35 checklists per floor in a high-rise building. Hence the quote.

After a few days of this regime, the curtain wall sub got it and started working to a schedule that matched others.

What's the point? Checklists should be natural part of doing a job. They are all about what to consider, and were not originally designed to be slavishly filled in. But there are times when that is what you have to do.

What are the principles & best practices? Construction needs to be scheduled so that diligent consultants and superintendents can review work in progress before it is covered and becomes impossible to review. I say "diligent consultants and superintendents" on the assumption that contractor staff will review all installations before they are covered. For any given project, the local authorities will advise when they expect to be called; the consultants should as well. I use a list of Minimum Required Field Reviews, customized to the details of each project, to identify my expectations. I use the same list at the proposal stage to help identify my level of effort, hence fees:

A	C	RA	N/A	Minimum Required Field Review	Comments/ Conclusion/ Action - Minimum Chief review intervals
x		x		Foundations - Includes dampproofing, waterproofing, weather barrier under slab, below grade/slab insulation	Once per building or once per elevation
x	BE			Window/Air Tests - Typically 1% of windows are tested, min. 2 per phase, one early and 1 late in phase. The later test may include air testing of a typical completed unit. Coordinate with Consultant which bldg(s) have windows tested.	Divide between phases or buildings; approx. 1/2 earlier, 1/2 later
x	BE M/P E	x		Pre-drywall - Requires insulation in place, and as req'd, ADA sealant and gaskets, window sealant, poly v.b., f'stopping, ductwork sealed, other special conditions, etc.	Once/bldg -> townhouses; once/floor others. Contractor identify RA review requirements for pre-drywall
x	R			Roof - Requires roofing & flashing complete - important to conduct early in project to establish performance requirements.	R = independent roofing inspector if spec'd;
x	BE			Exterior Pre-cladding - Includes preparations for cladding on all exterior wall surfaces, i.e., moisture barrier and strapping measures in place. Often combined with other milestone reviews depending on supt. organization of work.	Once/bldg -> townhouses; once/floor others.
x	BE			Exterior Envelope - Includes substantial completion of all exterior surfaces - may be combined with Pre-occupancy if circumstances permit.	Once/bldg -> townhouses; once/floor others
x	M/P E			Chases - fire rating pre-board - For lower fire ratings (up to 1h), typically requires 3 of 4 sides of chases be drywalled and taped/sanded and 4th side be ready for sealing. Verify method of sealing 4th side as acceptable per code/local building authority.	Once/bldg -> townhouses; once/floor others. Contractor identify RA review requirements.
x	M/P E	x		Corridor ceiling drops - Ceiling areas insulated as required, taped with penetrations firestopped.	Once/bldg -> townhouses; once/floor other. Contractor identify RA review requirements.
x	M/P E	x		Suite ceiling drops - Ceiling areas insulated as required, taped with penetrations firestopped.	Once/bldg -> townhouses; once/floor other. Contractor identify RA review requirements.
x	M/P E	x		Ext. wall f'stop - before v/b - Review before insulation in place (so penetrations are visible) – penetrations fire stopped and smoke sealed. Review after communication and security wiring completed.	Once/bldg -> townhouses; once/floor other. Contractor identify RA review requirements.
x	M/P E	x		Interior wall f'stop - Penetrations fire stopped and smoke sealed. Review after communication and security wiring completed.	Once/bldg -> townhouses; once/floor other. Contractor identify RA review requirements.
x	S M/P E			Pre-Occupancy - Bldg. complete to allow issuance of assurance letters. Deficiencies minor if any and seasonal in nature. Assumes later pre-occupancy reviews pick up these deficiencies on earlier buildings. Assurance letters are annotated to indicate these minor seasonal deficiencies.	Varies with each project

A = Designer; C = Consultant (discipline(s) named - BE= Bldg. Envelope; S=Structural; M/P=Mechanical/Plumbing; E=Electrical, others in Comments column); RA = Regulatory Agency; N/A = Not Applicable at this time; Item = Item to be considered by Designer, Regulatory Agency and/or Client; Comments = by D, B, RA or others. Note: ALL construction is to be reviewed by the Contractor.
Insert the following notations as appropriate: NC = Non-Conformance; Def = Deficiency; P= Photo attached; FR = Field Review required; RFI = Request for Information required; xx = not equired of this party; 0 = information required; x = complete/signed off.

Figure 177 - List of Minimum Required Field Reviews - Edited for space

RECORD the journey	Identify required field reviews	**Summary Principles & Tools**	
RESOLVE the issues	Use various techniques to ensure required reviews are provided	**i-WorkPlan (i-WP)**	
REVIEW the results	Review for conformance to identified criteria	**i-Actions (i-A)**	Edit a standard list(s) to suit the project
REMEMBER & learn to improve	Use the list to assist fee calculation	**i-KnowHow (i-KH)**	

Links: I arbitrarily deleted a few rows from the Minimum Required Field Reviews list above so that it would fit in the illustration space available. To see the full list, go to http://bit.ly/aagc59-3

For a sample of the curtain wall checklist that you do NOT want to fill out for each room, go to http://bit.ly/aagc59-2

60

Checking:
ITP - The High-Powered Checklist

"It doesn't matter what we think is the best format for our own checklists. This industrial client has their own format and that's what we have to use" – Experienced quality manager to stubborn quality director.

What's the point? Our quality manager for the industrial project was patiently explaining to me why we could not use the checklist format discussed in the previous tales in this book. I had already been told that it usually took months to have a project's i-WP approved by this client, involving many meetings and multiple revisions. It seemed I had my work cut out for me.

In the world of industrial building projects such as complex process manufacturing plants or structures auxiliary to major engineering works like dams, the designer's role, especially the architect's, may seem limited. Yet each plant needs buildings to house equipment, control centers, etc. Many architectural firms do satisfying work at high fees in this arena. Some years ago, an architect patiently explained to me that his firm designed all of the electric utility's structures. The overall value of these structures compared to the cost of a dam was minor, so the client did not care that the architect had fun with their design, increased the value of the architect-designed structures by 10-20%, and charged a 6% fee for architectural services – the architectural construction costs and fees were the equivalent of a rounding error in the overall project costs.

What are the principles? The downside of these interesting architectural opportunities is that the client may expect the designer to adhere to a complex and exhaustive quality management structure. The list of required field reviews for such an industrial opportunity may alternatively be called an ITP – Inspection and Test Plan. An ITP is generally much more detailed than the checklists we have discussed in previous tales. An ITP will include the approximate construction sequencing of sometimes complex work, with multiple inspection and/or hold points (a hold point is where you hold your breath, hoping the client will approve work to that point.). To respond to this, we developed a generic i-A for the industrial client mentioned in this tale that looks something like this (don't worry about reading it, just notice the complexity):

Item: C6 ITP - Inspection & Test Plan			Referred By					
Type: F6 GENERAL FIELD REVIEW			Action Number: A80033 R0			Priority: 11 Reference		
Status: Open			Subtype			Flamed:		
Identifier			Due			Opened: 2014/02/12	Closed	
Non Responsible:			Originator					

Supplier/Sub-supplier Name:		LOCATION Address, City & Country				Purchaser P.O. #:				
Supplier/Sub-supplier Work Spec:										
Activity No.	Activity Description	Location	Control Procedure Ref.	Responsible Party	Acceptance Criteria	Verifying Documents	Inspection Points			
							D	I	C	E
1	2	3	4	5	6	7	8	9	10	11
	APPROVALS/SUBMITTALS									
	[List any approvals required before starting activities]									
	ITP approved by Client – Consultant									
	Non Conformance dispositions completed and logged.									
	MATERIAL VERIFICATION									
	[List receiving inspections required]									
	INSTALLATION									
	[List Key installation Activities by Discipline]									
	TURNOVER									
	As-builts recorded and maintained									
	Final Documentation review completed									
	[List any other requirements for Turnover by Contract]									

LEGEND: S = Supplier I = Purchaser C = Authorized Inspector E = Owner W = Witness Point H = Hold Point M = Monitor R = Review

| Prepared by Supplier/Sub-supplier Rep. | | | Reviewed & Approved by Supplier/Sub-supplier Quality Rep. | | | Approved by Purchaser | | Reviewed by Owner | |
| Print Name/ Initial | | Date | Print Name/ Initial | | Date | Print Name/ Initial | Date | Print Name/ Initial | Date |

Figure 178 - ITP as i-A

Why bother? In conventional building construction, a deficiency that goes unnoticed will likely cause minor water leakage, a crack in the drywall, etc. I am not making light of building deficiencies, but with rare exceptions they are not catastrophic.

By comparison, in the industrial world, a deficiency may result in leakage of huge volumes of sometimes-toxic materials. Worse, it may cause an explosion or similar deadly result. For these and similar reasons, inspection in the industrial world is generally more continuous and rigorous than in building construction.

What are the best practices? As a result, industrial world checklists have evolved into much more complex documents with these typical characteristics for each step in the design, fabrication, delivery, installation and commissioning processes:

Acceptance Standards include steps for most or all activities on the ITP. In addition to things designers and builders would understand, such as structural steel erection standards, industrial checklists include items such as welding standards, piping fabrication and installation standards, etc.

Recording Standards may be as simple as an indication of review frequency; more likely, as complex as sophisticated tests that need to be completed and verified completely before next steps.

Hold Points are stages in industrial assembly where work stops until a specific review or test has occurred, been witnessed, passed and recorded.

Where possible, try to convince the client to adopt a simpler ITP format based on Tale #56. That tale identifies the importance of identifying "pause points" to separate portions of a larger checklist into more manageable segments – ITP hold points by any other name.

A pause point is a natural time to pause and consider that certain steps have been completed. In the medical world, pause points have names like "Before administering anesthesia", "Before making incision", etc. In building construction the natural pause points are "Before starting construction", "During construction", "During construction before each Review", "Before closeout" – it's probably no coincidence that construction "closeout" is like surgery "close up!"

If you are unable to convince your client or designer to use the simple approach of Tale #56, you will at least be armed with a clear understanding of checklist principles, and your lists will almost inevitably be shorter and clearer than they would otherwise be.

Conclusion: At the beginning of this short tale I mentioned being warned that the industrial client would take a long time and many revisions before they would approve our project quality plan and ITP structure. As described above, we identified the particular quality-related requirements of this client (there were many), then created some new procedures and tagged them with the client name for future reuse. We blended those into a more typical i-WP (62 pages + checklists), submitted the draft and awaited the feared response.

After a couple of weeks we were invited to teleconference with the client, whose staff for this project were located in three different cities. The client's senior project manager opened the meeting:

"We need you to make some revisions to your draft plan", he said. I thought, here it comes!

"Can you please provide an up to date list of our project staff and yours, including consultants, and a project org chart? With those, we are ready to go."

That's it, I thought? What about the multiple revisions, the arguments over ITP format? There were none.

RECORD the journey	Use ITP or Gawande-style checklist	**Summary Principles & Tools**	
RESOLVE the issues	Multiple hold/ pause points and sign off's do this	**i-WorkPlan (i-WP)**	
REVIEW the results		**i-Actions (i-A)**	i-A in ITP format – base it on client's traditional standard or Gawande
REMEMBER & learn to improve	Every industrial client is very different – capture those differences for next time	**i-KnowHow (i-KH)**	

## 61

Completing the Project:
"Are we Done Yet?"

"The building envelope consultant identifies issues every time he comes on site, and we deal with them as quickly as we can. But it's nine months into a twelve month project and he has yet to close one item!" – Project manager to quality director.

What's the point? I had been asked to attend a regular OAC (Owner Architect Contractor) meeting because the project team thought I, as a former practicing consultant, might be able to help them better manage the consultants, especially the building envelope consultant, who was a reasonable guy except for his hesitancy about closing issues.

I came early so I could have my own look around and review some of the correspondence. Sure enough, the site appeared in good order, the schedule was being maintained, the number of issues was reasonable for the stage and scale of the project – all good.

However, I did notice the other project consultants were all delinquent in closing out issues, just not as bad as the envelope consultant. So I decided on an approach with the project manager and awaited the start of the meeting.

At the appropriate moment in the meeting, the project manager introduced me and asked if I had any new business to table. I had listened to a recitation of all of the items that were "open" but which the contractor had advised for some time had been completed. Noting this, I made a modest proposal introduced in the guise of avoiding downstream claims for extra on account of schedule delays arising from issues not being closed, hence preventing construction from being closed in. The mention of claims caught the attention of the owner's representative and prime consultant.

In that pre-tablet era, I suggested all consultants simply use the multipart Quality Report that we had developed in response to superintendents asking for some means of capturing consultant non-conformances and deficiencies:

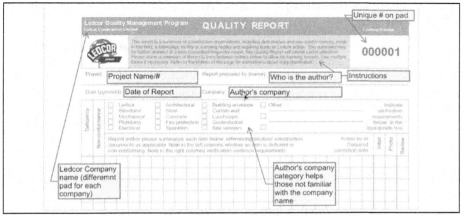

Figure 179 - Multipart Paper Quality Report

You will probably notice the "bones" of the field review report i-A from a previous Tale. In fact, the i-A followed after.

My suggestion caused an uproar, with no consultant interested in using a form with the construction company logo on it. But the opening allowed me to explain what a Quality Report is all about, emphasizing the importance of closing items raised. One after the other, the consultants, including the envelope consultant, volunteered that they would improve.

Two days after the site meeting, a list of items closed arrived from the envelope consultant, closely followed by similar lists from the other consultants. Did we care that they did not wish to use our form? Of course not. Interestingly, shortly after the site meeting two of the consultants modified their standard field review report forms such that they looked remarkably like our Quality Report. Imitation really is the best form of flattery!

What are the principles & best practices? Reflecting on the title of this tale, there are a handful of other key considerations that are broader than the state of any particular area as a project moves to completion.

Has each non-conformance, deficiency or other issue been resolved and closed out? A surprising number of consultants and clients do not well manage the very issues they have raised. Perhaps this is because the contractor remains responsible for the complete building construction regardless of how poor the record keeping of others may be. Many insurers and lawyers are of the opinion that in the event of a downstream problem arising from an issue without records of resolution during construction, the courts will look at each party's records and lean towards blaming parties with poor proof of what they did or did not do. Certainly you are more likely to end up in mediation or the courtroom if your construction records are a shambles. I always recommend to anyone I am counseling that they should keep their own records and they should be comprehensive and appear to be so. To paraphrase, "Issues should not only be resolved, they should be seen to be resolved."

Have weather related deficiencies been identified together with a schedule for their completion? Weather related deficiencies are those that cannot be completed because of the weather. For example, caulking is only practical in certain temperature ranges, which may be exceeded at times in winter or summer. Identify each weather related deficiency (there should not be too many) and attach to it a scheduled completion date, then issue reminders as the date approaches (automatic if you are using an i-A approach).

Has a deficiency holdback for each trade been identified and agreed to? A later tale discusses how to estimate deficiency holdbacks. If you are using i-Actions to identify each deficiency, you may be able to take advantage of the ability to attach a $ value to each deficiency. Look at the bottom right of the figure below.

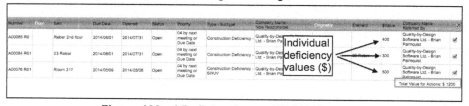

Figure 180 - 4 Deficiencies worth $1200 in total

Has each closeout submittal been submitted, reviewed and accepted? If you have been diligent earlier in the project in establishing an i-Action for each required submittal, including closeout submittals, it will be simple to track each one to completion.

D	B	C	N/A	Before closeout of this trade	Notes
x	x			Each non-conformance, deficiency or other issue of this trade been resolved and closed out?	All iT's should be closed
x	x			Any weather related deficiency been identified together with a schedule for its completion?	Open an iT for each deficiency
x	x			A deficiency holdback for this trade been identified and agreed to?	
x	x			Closeout submittals submitted, reviewed & accepted	Open an iT submittal for each closeout submittal

Figure 181 - Key closeout elements on any project

Note that designers will argue that it is not their responsibility to track submission of closeout submittals, only to review what is given to them. This is technically and contractually correct. However, designers who wish to keep their clients happy will work with builders to develop a system that works for both.

The figure below is from a more complex project and includes the typical range of closeout submittals that Consultants may need to look for:

Use this table to manage trade closeout. Recipient can provide details by selecting "Reply All", then adding information as appropriate and sending back.

D	B	C	N/A	Before closeout of this trade	Notes
x	x			Each non-conformance, deficiency or other issue of this trade been resolved and closed out?	All iT's should be closed
x	x			Any weather related deficiency been identified together with a schedule for its completion?	Open an iT for each deficiency
x	x			A deficiency holdback for this trade been identified and agreed to?	
x	x			Closeout submittals submitted, reviewed & accepted	Open an iT submittal for each closeout submittal
				[insert specific closeout submittal requirements in this column]	
	x			Trade contact info	
	x			As-Built drawings	
		x		Equipment list by system & supplier Info including local system support office	
		x		Certificates of Approval and Acceptance, including those of authorities having jurisdiction	
		x		Material & equipment technical specifications, productdata and catalogs	
	x			Operating instructions	
		x		Maintenance requirements, checklists & schedules	
		x		Trade maintenance after completion?	
		x		Warranty(ies) & Guarantee(s)	
		x		Extended warranty?	
		x		Spare parts list & contract quantities	
		x		Training details & list	Use iT's to manage training events
		x		Manuals	
		x		Videos	
		x		Demonstration schedule	

When the construction Phases of the work of this trade have been completed, consider the items above before closing out the work of this trade. Items not required should be marked with an "X" in the N/A column. During the course of closeout for this trade, keep Diary Notes with this Checklist describing progress. When the closeout Phase checklist, including iT

Figure 182 - Closeout Submittal list from Complex Project

Using Checklists to manage completion on smaller projects – I have been on small tenant improvement projects where the builder successfully used only one checklist for each trade, using the logical progression in the checklist to work through the simpler list of mockups, submittals, issues and deficiencies. Conversely, I have been involved in projects where there were hundreds of checklists and submittals.

RECORD the journey	Checklists	Summary Principles & Tools	
RESOLVE the issues	Use checklists, i-A's for significant issues	**i-WorkPlan** **(i-WP)**	
REVIEW the results		**i-Actions** **(i-A)**	i-A's for trade closeout
REMEMBER & learn to improve	Refine to match specific client/community needs	**i-KnowHow** **(i-KH)**	

Links: For a sample Trade closeout checklist, go to http://bit.ly/aagc61-1

## 62

Completing Pre-occupancy Field Review

"I can guarantee that the deficiency list I have prepared for these two suites will apply to the remaining 82. When you have completed all of these items in all 84 suites, give me a call." – Architect to Construction Manager

The client and contractor were in full panic mode. Occupancy showed on the contractor's schedule as occurring in two weeks, and we all knew the city required at least one of those weeks after field approval in order to complete their internal paperwork and issue an occupancy certificate. And if their review was not a "pass", they would decline to even return to the site for at least a week, presumably as punishment. Eighty-four purchasers' lawyers were anxiously awaiting the occupancy certificate as one important piece of the documentation that they would need to convey eighty-four titles.

The contractor and client had summoned me to the site. I had tried on the pre-inspection checklist discussed in a previous tale – the one where the contractor assures me everything is complete in the area to be inspected (in this case the entire building). The client then called me and begged me to come out "as a favor." Sigh!

The objective of my review was twofold: firstly, to assure myself that the architectural work scope was substantially complete so that I could sign my assurance letter; secondly, to determine if the project would be ready in time for me to call out the city inspectors in a week with some prospect of passing, so that I could then spend the next week chasing them all over city hall for their sign-off's.

The project was not looking good. I patiently worked through two suites at a rate of about ½ hour per suite. Even though I had reasonably automated means (for the day) of identifying deficiencies, it required ½ hour to: try every door, window and drawer; confirm penetrations through walls and ceilings were firestopped and escutcheon-plated; check finishes for nicks, chips, rips, gouges, etc.; ensure appliances were actually hooked up; yank each balcony rail while confirming all screws and bolts were in place; look for ragged/sharp flashing edges that might lacerate hands or worse, etc.; finally, annotate floor plans to capture the 50 or more deficiencies per suite. Simple math told me there were at least 4,000 deficiencies, probably more when the parkade, groundscape, etc., were considered. Even more with the plumbing, mechanical and electrical.

The deficiency lists were similar enough for the two suites that I had enough evidence. I handed the lists to the contractor with the comment at the beginning of this tale.

What's the point? At a certain point it is up to the contractor to just finish. We are all tired, perhaps frustrated and anxious, but it is for the contractor to just finish. An architect who painstakingly delineates hundreds or thousands of deficiencies is doing the contractor's job. While there may be times when this is requested by the client (all right so long as you have the time and are properly remunerated for your additional efforts), in general the contractor should already have done the 50-items-per-suite review previously and managed the trades in addressing that volume, so that when you arrive the list comprises, say, 2-3 items per room – at most.

What are the principles & best practices? Although every project is different, in many senses every project is the same. When a project approaches its completion there are predictable loose ends to tie up.

This can be a moment to savor the pending completion of a job well done, or a time for panic and problems.

D	B	C	N/A	Item to Review	Comments
				Things to do before you leave for the site:	
x				Contact the site superintendent to confirm that previously identified deficiencies have been attended to (if not, do NOT go)	
x				Ensure the site superintendent is available to tour the site with you	
x				Assemble plans and other construction documents, annotated to assist comprehensive review: ☐ all floor plans and roof plans ☐ supplementary sketches, site instructions, etc. ☐ finish schedule	
				Prior to performing pre-occupancy review:	
x				Remind the Contractor that inappropriate use of Consultant time may be charged as an additional service to the Client, thence may be recaptured from the Contractor	
x	x			Have Contractor confirm in writing that: ☐ the work is ready for occupancy ☐ inspections by the authorities having jurisdiction have been satisfactorily completed and evidence of such is available to the Consultant ☐ there is no ongoing work that might affect life safety such as completion of work in public areas	
x	x			Prepare a "punch" list or similar means of recording deficiencies, e.g.: ☐ reduced size floor plans that allow marking up with deficiency locations ☐ room finish schedule ☐ consider establishing a simple 1-letter coding for deficiencies such as: - A = adjust hardware - C = cleaning required - M = missing item - P = paint touchup required - R = replace damaged item - S = stain removal required - T = tighten hardware/connection (such a code will usually catch a high proportion of deficiencies of the type that)	
x				Bring to the pre-occupancy review the following equipment: ☐ Tape measure ☐ Camera ☐ indelible markers as well as conventional pens/pencils ☐ paper	
				Before starting the review:	
x				Explain to the Contractor the means by which deficiencies will be recorded during the review	
x				Explain to the Contractor that if, in the Consultant sole opinion the extent of deficiencies are too extensive, the review will be curtailed until deficienies have been properly attended to	

Figure 183 – Sample Pre-occupancy field review Checklist

This pre-occupancy checklist above balances practical considerations with reminders that the consultant can/should down tools if it appears the project is not near completion. Contractors prepare punch lists, consultants prepare deficiency lists.

To end this tale, know that many contractors have an amazing ability to complete work when all the excuses have run out. The client understood that if I organized a city inspection that was a failure, then he would lose at least one extra week – guaranteed (I put that in writing to cover myself). So I called the city and scheduled their inspection for a week hence. I further agreed, as an additional service for additional fees, to make daily reviews of portions of the building that the contractor deemed "ready", and to leave deficiency lists on site before leaving.

Miraculously, the work was completed on time, although I think paint was still drying in some areas as city inspectors worked through the building. Eight-four "deals" completed and we all lived to design and construct another day!

RECORD the journey	Check everything that a purchaser or tenant would	**Summary Principles & Tools**	
RESOLVE the issues	Identify typical issues and look for patterns	**i-WorkPlan (i-WP)**	
REVIEW the results	If there is a pattern of incomplete, identify it to the contractor & stop doing contractor work	**i-Actions (i-A)**	Can include suite or floor area plans
REMEMBER & learn to improve		**i-KnowHow (i-KH)**	

Links: For a copy of the pre-occupancy checklist, go to http://bit.ly/aagc62-1

## 63

Completing Substantially:
Be Clear when you are Done

"I'm sorry if my not releasing my assurance letters is holding up your occupancy, but the work's not done yet!" – Architect to Owner's rep

What's the point? Many things hinge on the successful and timely completion of a construction project: costly interim financing is supplanted by less costly and more effective long term financing; leases may be signed and sales deals closed; contractors get paid outstanding accounts; the warranty clock starts ticking; etc.

There is tremendous pressure on everyone to complete a project, but there is an equally dangerous liability for the consultant who says a project is done before it is. Although it has never happened to me, I am aware of a handful of architects whose reputations were destroyed when they certified completion before they should have; the municipality checked, found obvious areas incomplete and filed complaints with our professional association, which fined, reprimanded and suspended the over-eager architects.

What are the principles & best practices? In most jurisdictions there are two completion circumstances that are often confused. Substantial completion or substantial performance occurs when the jurisdiction's legislation or construction contract says it has (Great! you say, that was a big help!).

Most jurisdictions have one or more formulae that determine if substantial completion has been achieved – usually a high percentage of the contract value has been certified by the consultant/payment certifier.

OR the more pervasive and logical indicator of substantial completion is the answer to this simple question: "Is the building ready for occupancy for the purpose intended?" Several of my projects have had completion delayed by that principle regardless how much money has been certified.

A simple real life example from my practice: a multistory, multi-family residential building consists of apartments opening onto a central corridor with a central elevator and fire exits either end – a very common configuration.

The builder says the project is substantially complete. You review and indeed, all of the apartments are complete. But you notice there is no carpet in one of the exit stairs.

"We'll install that after occupancy, so the folks can move in at month's end, the end of the week."

I don't think so.

Problem is, as soon as the carpet installers start on the stairs, they will be blocking one of the two exits, rendering fire exiting unsafe.

It frequently happens that in the push to complete habitable, salable, rentable space, the access to and safe exiting from that space is left to the end.

At the more sophisticated end of this spectrum, in airport projects, areas are identified as ready for occupancy based upon evacuation zones. The 24/7/365 nature of these facilities requires that additions and renovations not interfere with safe use and egress from anywhere else in the airport at any time. With complex baggage handing, mechanical systems and security requirements, this can be very challenging.

A second, subtler issue that happened on the same residential building introduced above: *"We are all done, except the balcony rails on two suites were measured wrong and are in transit."*

In this case, the best course of action is to first, verify the extent of the issue – in this case all was otherwise ready in the two suites except for the guardrails.

With that knowledge in hand I went to city hall, explained the situation and asked if we could apply for occupancy "except not suites 208 and 209." City hall was more than happy to oblige, with the proviso that the builder install a sign on the two suites' entry doors indicating they were a construction zone, requiring adherence to proper safety practices. Easily done.

Occupancy was achieved after the stairs were completed and everyone except for occupants of those two suites was able to move in. The final two suites' rails arrived within the week, were installed, inspected by me and the guardrail engineer and we then applied and received occupancy for the remaining two suites.

By the way, all of the effort associated with additional field reviews, paperwork, visits to city hall, etc., was all additional services paid for by the client.

RECORD the journey	Be clear about occupancy vs. substantial	**Summary Principles & Tools**	
RESOLVE the issues	Use issue-specific i-A's to keep issues on the radar	**i-WorkPlan (i-WP)**	
REVIEW the results	Look throughout before confirming substantial performance	**i-Actions (i-A)**	Manage each issue as a high priority i-A
REMEMBER & learn to improve	Capture individual community occupancy requirements	**i-KnowHow (i-KH)**	

64

Completing:
Occupancy vs. Substantial

"We have substantial completion – why can't we move in?" – Owner to Architect

---

SUBSTANTIAL PERFORMANCE OF THE WORK

Substantial Performance of the Work is as defined in the lien legislation applicable to the *Place of the Work*. If such legislation is not in force or does not contain such definition, *Substantial Performance of the Work* shall have been reached when the *Work* is ready for use or is being used for the purpose intended and is so certified by the *Managing Consultant*.

Figure 184 - One Jurisdiction's Definition of Substantial Performance

What's the principle? Many design and construction team members get confused that the substantial performance of a construction contract (a.k.a. substantial completion) means a building is ready to occupy. Actually, a building is only ready to occupy when the municipality it sits in says it is. It's as simple and as complex as that. The municipality is not and does not want to be a party to the construction contract that identifies when a project is substantially completed or performed.

There are so many variations and local definitions that you should read the construction contract again to refresh yourself about the applicable method of determining substantial completion/performance for each project.

Perhaps because the municipality in which a building sits is responsible for its fire protection, for providing its utility engineering services, road access to it and a host of other municipal services, local government reserves the right to decide when it is ready for its citizens to move in.

"CERTIFICATION: I hereby certify that I am the owner or am acting on behalf of the owner and I acknowledge that before an Occupancy Permit will be issued, I must, prior to the proposed occupancy date, deliver to the City of Vancouver, Licences and Inspections Department, the appropriate Letters of Assurance as required pursuant to the Building By-law certifying that the building or applicable portion, for which an Occupancy Permit is being applied for, substantially conforms, in all material respects to the approved plans and the requirements of the Building By-law must be substantially complied with, before occupancy will be authorized for that date". An extra charge may be levied for any re-inspection of work after the "proposed occupancy date".

Figure 185 - One Jurisdiction's Partial Requirements for Occupancy

As a sidebar, read the indemnities included in the fine print of the occupancy permit application for your next project and ask yourself whether you or your client should sign this form. A portion of a sample application form is printed above. In almost every case after you have read the fine print on an occupancy application, you will decide you as a professional do not want to sign it. Although the wording above is relatively mild, I have always taken such forms to the client for signature. I may deliver an application as a courtesy to my client, or as part of my services, but delivery is not signing.

My most memorable occupancy permit application experience was in a suburb of my home city, Vancouver. I was under some pressure by the client to apply for the permit – their project manager was on holiday. When I read the fine print, it declared that the signatory to the application agreed to be responsible for any future problems with the city services in the street right of way bordering the project. Say what!

When I called the client to explain why I could not sign the application form, his response was "You are the first professional ever to refuse to sign this form on our behalf." I nonetheless refused (politely) to sign, the developer did and all proceeded to a successful conclusion – and the developer continued to hire me for projects in that and several other municipalities.

Perhaps he was impressed that I had actually read the fine print.

RECORD the journey	Be clear about local occupancy requirements	Summary Principles & Tools	
RESOLVE the issues	Use i-A's to manage emergent issues	i-WorkPlan (i-WP)	
REVIEW the results	Look to others to sign application forms	i-Actions (i-A)	Use i-A's to manage emergent issues
REMEMBER & learn to improve	Capture specific community requirements	i-KnowHow (i-KH)	

## 65

Completing:
How Much to Holdback?

"You're charging a $200 deficiency holdback against the door handle! It's only worth $50!" – Installer to Architect

"What do you need a deficiency holdback for? You already have 10% holdback, isn't that enough?" – Contractor to Architect

What's the point? There are more formulas for calculating holdback than there are consultants on a project. Several are equally valid.

The point of a holdback is in its name – it's an amount of money to hold back until a deficiency has been completed. That's it.

As to calculating how much is an appropriate amount, here are two of the formulas I have heard of:
- Double or triple the actual cost to complete or fix.
- Cost to complete plus $100.

However you figure it, you need to have enough money set aside so that if the relevant contractor decides not to complete the work, the client can hire someone else to complete the work. In fact, that's how I calculate holdback.

If the door hardware supplier refuses to complete the supply and installation of missing hardware, how much will your client spend to hire a replacement contractor? Or how much will the client pay another mason to finish off the fifty deficiencies the original mason left unfixed?

There are no widely accepted formulae for deficiency calculation, so do what suits you. Just make sure it's logical and you are consistent.

Total Items: 4

Identifier	Number	Item	Opened	Type - Subtype	Company Name - Now Responsible	Originator	Company Name - Referred By
	A00085 R01	Rebar 2nd floor	2014/07/31	Construction Deficiency -	Copyright 2014 Brian Palmquist - Brian Palmquist		Copyright 2014 Brian Palmquist - Brian Palmquist
	A00084 R01	03 Rebar	2014/07/31	Construction Deficiency -	Copyright 2014 Brian Palmquist - Brian Palmquist		Copyright 2014 Brian Palmquist - Brian Palmquist
	A00083 R0	03 Rebar	2014/07/31	Construction Deficiency -	Copyright 2014 Brian Palmquist - Brian Palmquist		Copyright 2014 Brian Palmquist - Brian Palmquist
	A00076 R01	Room 217	2014/05/08	Construction Deficiency - SWJV	Copyright 2014 Brian Palmquist - Brian Palmquist		Copyright 2014 Brian Palmquist - Brian Palmquist

Total Value for Actions: $ 1200

Figure 186 - i-A Total $ Value Noted

What are the principles & best practices? Deficiency holdback is often confused with what is called in my province "builder lien holdback." While the deficiency hold-back is for incomplete work, builder lien holdback, a.k.a. construction lien holdback, mechanic's lien holdback, etc., is for the sole purpose of guarding against a contractor not paying a subcontractor, or a subcontractor not paying a supplier or worker. In Tale #46 I talk about a contractor who filed for bankruptcy and skipped town, leaving money owing to many of his subcontractors and suppliers. The builder lien holdback, 10% of the certified cost of construction in this case, was used to pay at least a portion of accounts owing.

Most provincial or state jurisdictions will have legislation and regulations designed to protect those unable to afford a lawyer to press a claim for payment. Do not confuse deficiency holdback with builder/mechanic lien holdback and never blend them togeth-er, no matter what suggestions are made.

The most typical builder lien holdback is 10%, meaning that 1/10th of each payment from the owner to the contractor is "held back" and usually kept in a separate bank account, certainly not the contractor's bank account. Your client's contract or the juris-diction's holdback percentage may differ, so check before assuming 10%.

After a project has been declared by its prime consultant to be substantially complete, any workers, suppliers and subcontractors who believe they are owed money have a defined period of time, usually called the "lien period", to make a simple filing in the court of the place of the work, indicating how much they believe they are owed. The day after the expiry of the lien period, lawyers usually representing the owner and con-tractor will search the legal title of the property to see if there are any liens. There being none, the contractor will fairly expect prompt payment of the lien holdback from the owner. This payment is NOT handled the same was as a monthly progress payment, rather, it is typically processed and paid within a short, prescribed number of days, typically ten days where I work.

If there are any liens, the lawyers for the owner and contractor will investigate the cir-cumstances. There is no requirements that the consultant or her/his lawyer be involved in this; if you are asked to be, first check with your insurer, then if you have a green light, charge additional hourly for your services.

Sometimes the contractor will simply pay out the lien claimant as a kind of "good rid-dance" gesture. If the contractor strongly disagrees with the lien, she/he may challenge the lien. This usually involves paying the amount claimed into the trust of the court, which "clears title" so that other accounts can be settled, property can be conveyed, etc. The parties may then become involved in litigation without delaying any other pay-ments to Trades. It's usually the claimant who has to initiate legal action within a pre-scribed time period, often one year. Meanwhile the balance of the lien holdback will be paid to the contractor for her/his other final payments.

Hopefully you see the difference between these two types of holdback: the deficiency holdback ensures deficiencies will be completed; the lien holdback protects against non-payment of subcontractors, suppliers and workers. Don't confuse them!

RECORD the journey	Identify deficiency holdbacks early	Summary Principles & Tools	
RESOLVE the issues	Recommend release of lien holdbacks when advised by lawyers	i-WorkPlan (i-WP)	
REVIEW the results	Review for conformance to identified criteria	i-Actions (i-A)	Use i-A's for records and issue resolution
REMEMBER & learn to improve		i-KnowHow (i-KH)	

66

Completing Commitments:
Assurance Letters

"The inspiration for the LoA [letters of assurance] system established in British Columbia was, to a large extent, the collapse of part of the roof structure at the Station Square Mall in Burnaby on April 23, 1988." – Elliot Lake Inquiry, p. 668[1]

"Letters of Assurance in specific instances ... document the parties responsible for design and field review of construction, and to obtain their professional assurances that the work substantially complies with the requirements of the BCBC 2006 [building code], except for construction safety aspects, and that the requisite field reviews have been completed. Construction safety is the responsibility of the Constructor." - Guide to the Letters of Assurance in the BC Building Code 2006[2]

"For the construction of any buildings requiring the services of more than one professional consultant, either a professional engineer or an architect should be designated by the owner or the owner's agent as the prime consultant to perform the roles and responsibilities of that position, as defined by one or the other or both of the Professional Engineers of Ontario (PEO) and the Ontario Association of Architects (OAA) ." – Recommendation 1.27, Elliot Lake Inquiry[3]

What's the point? In Tale #1, "Making Commitments", we introduced professional assurance letters (A and B in British Columbia, A1, A2, B1 and B2 in Alberta), which are executed and submitted at the building permit stage; no building permit will be issued until they are delivered. At project completion the loop is closed by having the same professionals provide assurance letters called "C's" (for "Complete"?) prior to occupancy.

What are the principles? Schedule C's essentially confirm that the commitments made in those earlier assurance "A" and "B" letters have happened, i.e., the project has been substantially constructed in compliance with the building code; no occupancy certificate will be issued until they are delivered.

---

1   Hon. Paul R. Belanger, Report of the Elliot Lake Commission of Inquiry, Queen's Printer for Ontario, 2014, Section II, Causes of the Collapse, http://www.elliotlakeinquiry.ca/report/Vol1_E/ELI_Vol1_Ch03_E.pdf p. 668
2   Building & Safety Standards Branch, Ministry of Public Safety & Solicitor General, Province of British Columbia, December 2010, Edition 5a, at http://bit.ly/1oDa01Q
3   Hon. Paul R. Belanger, p. 667

I hereby give assurance that

(a) I have fulfilled my obligations for coordination of field review of the registered professionals required for the project as outlined in Subsection 2.2.7, Division C of the British Columbia Building Code and in the previously submitted Schedule A, "CONFIRMATION OF COMMITMENT BY OWNER AND BY COORDINATING REGISTERED PROFESSIONAL,"

(b) I have coordinated the functional testing of the fire protection and life safety systems to ascertain that they substantially comply in all material respects with

   (i) the applicable requirements of the BC Building Code and other applicable enactments respecting safety, not including construction safety aspects, and

   (ii) the plans and supporting documents submitted in support of the application for the building permit,

(c) I am a registered professional as defined in the British Columbia Building Code.

Figure 187 - Schedule C-A Commitments

What are the best practices? The Coordinating Registered Professional (CRP) named at the start of the project in Schedule A executes a Schedule C-A confirming overall coordination duties have been completed, then usually a C-B along with other registered professionals confirming specific consultant duties (e.g., architectural) have also been completed:

(a) I have fulfilled my obligations for *field review* as outlined in Subsection 2.2.7, Division C of the British Columbia Building Code and in the previously submitted Schedule B, "ASSURANCE OF PROFESSIONAL DESIGN AND COMMITMENT FOR FIELD REVIEW", and

(b) those components of the project opposite my initials in Schedule B substantially comply in all material respects with

   (i) the applicable requirements of the B.C. Building Code and other applicable enactments respecting safety, not including construction safety aspects, and

   (ii) the plans and supporting documents submitted in support of the application for the *building* permit,

(c) I am a *registered professional of record* as defined in the British Columbia Building Code.

Figure 188 - Schedule C-B Commitments

I had some involvement in refinements to the wording of these schedules as part of later versions of the BC Building Code – at times a bit like arguing about how many angels can dance on the head of a pin. But I have no doubt lives (and reputations) have been saved as a result of such assurance letters. We should all have them.

RECORD the journey	A completion letter to match each original letter	Summary Principles & Tools	
RESOLVE the issues		i-WorkPlan (i-WP)	
REVIEW the results	Review letters for completion & accuracy	i-Actions (i-A)	For issue resolution
REMEMBER & learn to improve	Different assurance letters for different jurisdictions	i-KnowHow (i-KH)	

Links: BC Assurance letters are at http://bit.ly/aagc67-3 ; Alberta building code schedules are at http://bit.ly/aagc67-2

67

Completing:
Certificates - be Clear when you are Done

"You're serious? You want me to go all the way to the site just to post this certificate of completion?" – Project Architect to Principal Architect

"It would really help our sales program if you would date the completion certificate last Friday instead of to-day." – Owner's rep to Architect

What's the point? It was one of my first projects and I was proud how it turned out and that it had been completed on schedule. The Principal had accompanied me around the site and was equally pleased. When we returned to the office he completed, signed and sealed the certificate of completion, then asked me to return to the site to post it – hence my outburst in the quotation above.

What are the principles? Rather than reaming me out for impertinence, my Principal patiently explained that it was essential to physically post a certificate of completion on the site the same day as substantial completion was declared. *"Any worker or supplier who is owed money looks for that certificate, knowing that declaration of substantial completion starts the clock ticking on the filing of any claims against the project. Delaying the posting by one day reduces the claimant's rights by one day."*

What are the best practices? The importance of this simple task is such that it should only be delegated to someone you trust, such as an employee or associate of the firm. For good measure, my practice is to photograph the posted certificate in its context after posting. On or beside the front door of the completed building is a good place, but on the inside, visible through glass, so it can be easily seen but not easily removed.

Builders Lien Act

(Section 7(10))

Certification of Completion

I .... [name of payment certifier], of .... [address], British Columbia, certify that, for the purposes of the *Builders Lien Act*, the following contract or subcontract was completed on .... [month, day, year]:

Street address or other description of the land affected by the improvement:

Brief description of the improvement:

Brief description of the contract or subcontract, including the date of the contract and the names of the parties to it:

Signed: .... [signature of payment certifier]

Dated .... [month, day, year]

Figure 189 - Typical Certificate of Completion[1]

---

1 Builder Lien Act, "Builders Lien Forms Regulation", B.C. Reg. 1/98 O.C. 12/98, Queens Printer, Victoria, B.C., http://www.bclaws.ca/Recon/document/ID/freeside/10_1_98

The sample above is from the province where I live. Most jurisdictions will have minor variations on the theme. Essential elements include:

The name of the payment certifier: In most jurisdictions it is part of the duties of the person who recommends payments to the contractor to prepare (and post) the certificate of completion.

The date of completion: This is the date of substantial completion/performance, NOT the date of occupancy. This is the date from which the lien filing period starts. It will be different in different jurisdictions, so should be checked each time and diarized.

Address and brief description of improvements: When it comes to litigation affecting property, courts work almost entirely by addresses. They do not care about the name of the building, but to avoid confusion they expect a simple description such as "residential building", "commercial building", "airport structure", etc.

Who are the Contractor and Client? Noting that any subsequent lien or other litigation will involve either or both of the contractor and client, the courts would like this information upfront.

RECORD the journey	Post completion letters promptly & prominently	**Summary Principles & Tools**
RESOLVE the issues	Refer to the legislation	**i-WorkPlan (i-WP)**
REVIEW the results		**i-Actions (i-A)**
REMEMBER & learn to improve	Capture jurisdictional differences re timing, etc.	**i-KnowHow (i-KH)**

Links: British Columbia Notice of Certification of Completion at http://bit.ly/aagc68-1; British Columbia Builder Lien Act at http://bit.ly/aagc68-2

## 68

Warranty:
Why is it Always Different?

"We've used the same cleaning products on the same flooring in our other care home finished two years ago, and there have been no problems." – Owner to Contractor and Architect

What's the point? By the end of most projects the client, consultants and contractor are all tired of the project, ready to move on to the next adventure, perhaps after a bit of vacation time. It's easy to forget that pesky warranty, which usually requires the contractor to repair a variety of conditions during the warranty period (usually one year). For example:

If the wood-frame building shrinks, small cracks will need to be filled and painted over by the contractor's forces.

If a mechanical fan fails after only six months, it needs to be replaced by the contractor's forces.

I like to think of warranty items as deficiencies that only become evident after the building is occupied. As a general statement, the contractor remains responsible for the duration of the warranty for the performance of the building and any costs associated with repairing defects arising from construction.

What are the principles & best practices? As described in the typical construction contract below, the warranty process generally includes these steps:

The owner gives the contractor notice in writing of observed defects.

The contractor repairs the defects and then advises the owner, for re-review.

Although most contracts require the contractor to correct defects "promptly," as a practical matter only serious items such as building envelope or system leakage are addressed right away, due to requirements for tenant notification, access, etc.

Usually the contractor will collect non-serious defects (e.g., cracks in drywall due to settlement, malfunctioning window hardware, etc.) during the course of the warranty period. Late in the warranty period (say month 11 of a 1-year warranty), a walk through will be arranged involving the contractor, owner and consultant(s) as required by the nature of the defects. This on site meeting gives the contractor the opportunity to argue if a defect appears to arise from move in damages, occupant wear and tear, etc.

Interestingly, there is generally no specific requirement for consultants to attend or participate in the warranty process in many standard client/consultant agreements. But as noted below in clause 12.3.3 of a standard construction contract, involvement in warranty is one of the 183 places (in this contract) where the consultant becomes involved through the requirement that the *"...Owner, through the Consultant, shall promptly give the Contractor Notice in Writing of observed defects and deficiencies which occur during the one year warranty period."*

In practice, many clients expect their consultants to participate in warranty matters. Because this is a service not included in most client/consultant agreements, it is an additional service for which a consultant may choose to charge an additional fee. I seldom have, believing client goodwill and the learning experience to be reward enough.

GC 12.3 WARRANTY

12.3.1   Except for extended warranties as described in paragraph 12.3.6, the warranty period under the *Contract* is one year from the date of *Substantial Performance of the Work*.

12.3.2   The *Contractor* shall be responsible for the proper performance of the *Work* to the extent that the design and *Contract Documents* permit such performance.

12.3.3   The *Owner*, through the *Consultant*, shall promptly give the *Contractor Notice in Writing* of observed defects and deficiencies which occur during the one year warranty period.

12.3.4   Subject to paragraph 12.3.2, the *Contractor* shall correct promptly, at the *Contractor's* expense, defects or deficiencies in the *Work* which appear prior to and during the one year warranty period.

12.3.5   The *Contractor* shall correct or pay for damage resulting from corrections made under the requirements of paragraph 12.3.4.

12.3.6   Any extended warranties required beyond the one year warranty period as described in paragraph 12.3.1, shall be as specified in the *Contract Documents*. Extended warranties shall be issued by the warrantor to the benefit of the *Owner*. The *Contractor's* responsibility with respect to extended warranties shall be limited to obtaining any such extended warranties from the warrantor. The obligations under such extended warranties are solely the responsibilities of the warrantor.

Figure 190 - Typical Construction Warranty Provisions[1]

Warranty notices can arise from a number of sources: the owner; property manager; tenant or suite owner; etc. Calls may come to any staff member at the consultant, the owner or the contractor. This can cause the creation of a wide variety of confusing warranty records.

My preference (of course) is to open an i-A for each warranty item, which can be referred to the contractor. If it is dealt with and closed before the end of the warranty period, it will recede from view but remain in the project database. Otherwise it will sit quietly in the database until it is collected with its cousins near the expiry date of the warranty. The determination of action required can be recorded with the i-A, as well as progress, completion and sign-off.

Because I think of warranty notices as deficiencies that have arisen after completion, I use a construction deficiency notice to capture them. As with the deficiency notices mentioned in an earlier tale, your practice may cause you to edit this sample:

---

1   Canadian Construction Documents Committee, CCDC 2, AGREEMENT BETWEEN OWNER AND CONTRACTOR, p.30

Item	Instructions	Record
Description of deficient condition	Summary	
Originating Inspection Report	Reference # and issuing company	
Document reference	Drawing/sketch #, spec section, etc.	
Location	level, grid coordinates, elevation, etc.	
Immediate Action(s) required (if any)	as noted by Consultant/Inspector or Contractor personnel	
Other work affected by this deficiency	*(See Note 1 below)*	
Trade sign-off (where required, sign at right)	Indicate Company and print name	
Approval Sign-Offs:		
Contractor	Name:	Signature:
Consultant	Name:	Signature:
Owner	Name:	Signature:

Figure 191 - Warranty Notices

Latent Deficiencies – a special type: When drafting AAGC, I initially included as a sample deficiency in this tale, "If a sheet flooring material shifts and settles for no apparent reason, it needs to be replaced." Upon review, I moved it here as an example of a deficiency that only becomes apparent after the expiry of the contracted warranty. This actually happened on one of my first projects as a junior architect:

Shortly after the expiry of the one-year warranty, the residential care home's sheet flooring began to settle in some places and bubble in others. Investigation exonerated the cleaning staff, their products and procedures – in fact, they were using the identical cleaning products on the identical sheet goods in two facilities built before this one, with no apparent issues.

Investigation also revealed that the contractor had substituted a different filler and a different adhesive in parts of the building without advising the architect (read me) or the owner. The cost estimate to fix the flooring was significant enough that the contractor retained a materials science specialist, who concluded the substituted adhesive and filler were incompatible with each other. I thought we were in the clear.

BUT the materials scientist also concluded that at least some of the settlement and bubbling occurred because the plywood subfloor we had specified was of the wrong grade – it was sheathing grade, should have been finish grade for better adhesion of the sheet goods. *"But,"* we said, *"We had a tight budget and were trying to save you [the client] some money!"* Guess what? We shared the cost of repair with the contractor, on advice from our insurer.

The only good news from this story was that the client was sufficiently impressed with our "stepping up" that we were hired for his next project – and never again used sheathing grade plywood under sheet goods!

Most jurisdictions provide that an injured party has a year or more from the time of discovery to register such an issue. Your time line may differ.

RECORD the journey	A notice for each Warranty item	**Summary Principles & Tools**	
RESOLVE the issues	Each item requires contractor action	**i-WorkPlan (i-WP)**	
REVIEW the results	Participate in warranty review	**i-Actions (i-A)**	Warranty item detail
REMEMBER & learn to improve	Emergent warranty items	**i-KnowHow (i-KH)**	

Links: To access the warranty notice, go to http://bit.ly/ZRzeOj

## 69

Organizing:
Evaluating What's out There

"Every day I turn around and there is a new app saying it will solve all my problems – how can I make a decision?" – Just about anyone

What's the Point? Almost every week one of my colleagues breathlessly brings to me the solution to all the world's design and construction ills. Part of my current job is to evaluate these offerings to determine which, if any, are worthy of exploration. I have to be objective and fair. Easier said than done.

What are the principles & best practices? In order to objectively consider what's on offer this week, I have developed an evaluation table. Being essentially lazy, I generally send the table to the proponent and ask she/he to fill it in, preferably in consultation with the excited software vendor.

This table includes all of the key software performance factors I have identified over 10+ years, organized around the three toolsets I introduced at the beginning of AAGC..

For purposes of this e-book, I have reproduced the table, annotating the "Comments" column in italics to provide further explanation:

#	Item	Comments
1	**i-WorkPlan**	Does the i-WP or similar capability even exist? *Without an i-WP you are managing a bunch of drawings and documents, not a project.*
.1	instantaneous	Web-based; an i-WP can be prepared quickly – *it should take hours, not days to draft an i-WP*
.2	interactive	Is a Database; it can be printed out; many users on many projects can use it. *Without a database, every project is a standalone event and you cannot easily build on past experience.*
.3	individual	Includes "no access", read only, read write and administrator access as a minimum; access is adjustable on a project by project basis. *Sometimes it is important for an individual to have more access on Project A than on Project B.*
Instructions: Mark "Yes" or "No" where a capability is clear. Identify "?"'s where performance is unclear. N/A = Not Applicable.		

#	Item	Comments
.4	individualized	Standard content can be supplemented by content customized to client, consultant, contract type, cost, contractor, climate, community, complexity, calendar, and construction type. *The "10C's" – it is essential you are able to tag content as applicable to, say, Acme Construction, Advantage Architects + municipality of Whatever.*
.5	instructive	Work instructions are included for each task and are easily accessible; updates are automatically deployed to all active projects. *If improvements and updates are not automatically deployed you will go mad. This is a fundamental element of ISO 9001*
.6	inclusive and immediate	Supporting forms and templates are attached to relevant tasks or otherwise easily accessible; updates to these are automatically deployed to all active projects. *Same reasons for auto-update as .5 above. Also if the correct forms are not immediately to hand they will not be used.*
.7	informative	Diary space is provided for each task in each project in the database; open/closed status is easily visible. *Replaces all of the individual diaries that are never available years later when needed.*
.8	issue oriented	Emerging issues can be associated with standard tasks. *Issues should not "float" separate from the procedures or building parts they relate to.*
.9	incomplete	i-WP can be updated "on the fly" as circumstances change. *Things change during the course of a project – buildings are added, and floors, etc.*
2	i-Action	
.1	integrated	Issues, actions, open and completed forms are attached to i-WP tasks; updated templates and forms are automatically deployed to current projects; items can also be organized and reported through a variety of filters such as "assigned to", assigned by", due date, etc.; replies to items issued via email are automatically captured to the originating item. *This reporting & filtering replaces all of the extra spreadsheets & lists that otherwise plague your life.*

Instructions: Mark "Yes" or "No" where a capability is clear. Identify "?"'s where performance is unclear. N/A = Not Applicable.

#	Item	Comments
.2	immediate	Items are assignable by immediate email in office and in the field. You want to get the i-A on its way as soon as possible. *And a record should be automatically created.*
.3	identifiable	Issued items have a distinctive and informative appearance. *To distinguish from the morass of email.*
.4	individual	Items are assignable to an individual; can be reassigned as work flow progresses; attachments are collected so visible to all; assignee and assignor are reminded of due date; a user can see all items assigned to/by him/her. *Workflow essential, also ability to filter to create individualized "to do" lists.*
.5	inclusive	Items have attached forms and templates where applicable, e.g., RFI, Submittal review. *If the correct forms are not immediately to hand they will not be used.*
3	**iKnowHow**	
.1	integrated	Part of the same database as the i-WP and i-A's; i-A's are easily imported into the knowledge database; knowledge items are easily exported into other projects; provides a single list of users and contacts with associated information such as CV's, experience records. *If it's not easy to capture it won't be.*
.2	immediate	New knowledge leading to i-WP or i-A revisions is immediately deployed to current projects.
.3	intelligent	Simple Subject Matter Expert (SME) review process. *If it's hard to get proposed new knowledge into the hands of SME's, and hard for them to review it, it will not happen.*
.4	investigative	Can use experience database to ask questions to staff with specific expertise; a continuous archive is created and available; drafting, issue, reissue, & refinement of contents are automatically recorded & accessible. *Experience + Expertise needs to be integrated with the knowledge database.*
.5	inclusive	Expertise database is maintained by users, not centrally. *This is a personal preference. People manage themselves more currently and accurately than their supervisors do.*

Instructions: Mark "Yes" or "No" where a capability is clear. Identify "?"'s where performance is unclear. N/A = Not Applicable.

#	Item	Comments
4	Other	
.1	Cost	Cost effective solution. *Pushiness is required for evaluators – costs are all over the map & most vendors are cagey.*
.2	Compatibility	With which tablets and smart phones? iPad? Android? Windows? *Most software will only get one or two out of three.*
.3	Training	instructions are available online and logically organized.
.4	Storage	Secure and backed up; synchronized between office and field. *The cheap apps (think $15.99 or less) usually don't have this and it is essential.*
.5	Old School	Supports form/template print out and fill in by hand.
.6	Cross Platform	Can the application import/export data from other applications? Which? Easy?
Instructions: Mark "Yes" or "No" where a capability is clear. Identify "?"'s where performance is unclear. N/A = Not Applicable.		

Figure 192 - Software Evaluation Table

## 70

"Wisdom" from the Trenches - 10 Rules

"Dear Contractor: You may be right; there may not be a good reason why I designed it this way, BUT please ask before you change it!" – Real notation on a young architect's title block (not mine).

"You have to have 10 rules at the end – after all those tales the audience deserves a reward!" – Spouse to Author/Architect

Looking back on the preceding 69 tales, I believe they can be summarized in the following ten rules that should guide the construction phase work of designers, builders and owners:

#1 – Organize the practice and the project will follow – Each building is unique, but the process of construction is predictable. Much of the experience gained on a project may be applied to future projects, if and only if you organize your practice/business to start with strong basics, then capture new knowledge and integrate it with your core professionalism. Taking this approach will make you increasingly efficient and foster a "right first time" mindset that will make you more profitable.

#2 – Keep the complex simple – The three toolsets and four principles recommended to manage design and construction are really all that you need. What does not fit in them is probably unnecessary, certainly not separate logs, lists and other duplicate or transcribed writings. The emergence of smartphones, tablets and web-based software fosters efficiency in all construction processes when a "keep it simple" approach is taken.

#3 – Understand what has been agreed to – Insist on reading all consultant and construction agreements, including your own. Raise your hand and ask as many questions as needed to be clear about your own and others' roles, responsibilities and scopes of service.

#4 – Help others understand what they have agreed to do - Ensure other members of the team, especially those you are charged with managing, fulfill their scope.

#5 – Begin as you mean to continue – Ensure you are prepared for the initial site meeting, payment certification, submittals and field review. Know your agenda and the full extent of your review responsibilities.

#6 – Share the work, especially when it changes – Value engineering, alternatives and substitutions are just a few examples of potential changes to the design you have worked so hard on. Politely insist that the builder, owner and other consultants participate – do not do the heavy lifting for them. And where your design is about to change, get paid to make the changes, including verifying that your design remains in the public interest.

#7 – Work with the builder – Contractors want the same as you, a well built design that is a credit to the entire construction team and profitable for all. Understand how you can assist contractor field staff with their challenges without doing the contractor's job – you will save lots of your own time and grow in the esteem of your client and the builder.

#8 – Work with the Community – Ensure the planning and building authorities and other regulatory agencies are "kept in the loop." Meet with them onsite, listen to their priorities and help them achieve their objectives. Your community reputation will be enhanced.

#9 – Review the construction of your design – Identify and implement a level of field review that will maximize the likelihood that your design is built as you would wish. Make clear to the builder which elements are most important to you – she/he will not necessarily know. Pay particular attention to those elements of your design that you know make it stand out, while not ignoring the balance of the design, especially measures that are in the public interest for a safe, secure building.

#10 – End as you began – Be clear about what needs to happen for you to certify that the work you designed has been properly constructed and is ready to occupy. Help the owner and builder to complete and occupy, but always with the public interest foremost in your mind. Pay attention to loose ends such as weather-affected deficiencies. Help where possible with post-construction warranty items. Remember that these final impressions of your services by your clients, builders, other consultants and the broader community will be their lasting impression of you.

The appendices that follow suggest ways the readers of AAGC can share their ideas and experiences with the broader audience of readers, and include a glossary of terms.

"Content or Conversation - Pick One"

About a week before writing this conclusion, I was contacted by a young, earnest former colleague who now lives in another city. I first met her when she worked on a major project that was using my Quality-Works.net (QW) software. She is now working on a startup, with a code writer in another city and an interface designer in yet a third. She was calling to tell me about a great new piece of software they had discovered for collaboration. It was fast, it enabled her team to come together virtually when they needed to with their unique contributions – all good. She called me about this application (not hers) because she thought it resembled QW, so I should look at it.

Ever curious, I signed up for a free demo and looked for the content – not her and her colleagues' content, which is naturally confidential, rather for the content framework that would help her team engage in the creative process. What parts of this software would help them: analyze a problem; research the issues; brainstorm solutions; synthesize to arrive at the best solution; design its details; "build" them in code; and capture new knowledge along the way, for use on the next projects - sound familiar?. I quickly discovered there was almost nothing there.

In fairness, the platform that she wanted me to investigate works at lightning speed – once it has your content. It shares your content with cool, "flat design" graphics (the latest web style) – once you have developed your content. It allows you to organize your thoughts - once you have developed them. But it has no process, little structure or content of its own. It is a true platform, that is to say a flat, empty space that only begins to gain substance when populated with the content of others. In the case of design and construction, it's as if this was a building platform or slab on grade hosting a number of workers, conversing amongst themselves (likely on social media) amidst piles of building materials without design, construction and materials knowledge or direction in the form of a documented design, yet with the expectation that from this there would somehow arise a building. I was surprised that she would think it was anything like my software.

As I began to ponder this apparent contradiction, I realized that I was to blame for her misconception. Well, not entirely me, but certainly my generation. Every day one reads about the latest app that calls you a cab, manages your shopping, or shares your thoughts whether profound or fatuous. Perhaps these apps have arisen in the absence of successful capture and communication of the substantive professional content needed to design and build a real building. In fact, where is the application that instructs you how to design and build a building? Is there an app for that?

The answer, of course, is "No!" There isn't an app to replace a generation of experience and a bit of wisdom. But I truly believe we can do better than an app. The fact that we can access many apps that do one or two things really well is not going to magically transfer knowledge between generations, just as the Xerox machine of my university days never managed to inject knowledge directly into my brain.

In the introduction to An Architect's Guide to Construction, I talked about three integrated toolsets that underlie design and construction: the i-WorkPlan that describes each step involved in a building's design or construction; the i-Actions that manage the small day to day design and construction steps, as well as emergent issues; and the i-KnowHow that captures new or refined knowledge "on the fly", constantly building the knowledge database of a practice so that there is substance and content to bequeath (or sell) to the succeeding generation. My i-WorkPlans come from a library of more than 400 procedures, supported by well over 100 i-Action templates. I'm not bragging, that's just the complexity of design and construction. And of course, over time your i-Know-How will become huge and valuable as mine has.

I challenge my older readers to start with their i-KnowHow. Assemble it as best you can. It is your greatest legacy even more so than the buildings you have designed and constructed. As you identify your knowledge, use it to infill the processes you use for design and construction. Use the "10C's" described in the introduction, or your own C's, to integrate the logic of different building types, locations, clients and collaborators into the procedures that underpin your i-WorkPlans and the details of your i-Action templates.

I have elected to share much of my own i-WorkPlan, i-Action and iKnowHow through the tales in this book and links to Quality-Works. I invite you to borrow what you will, challenge what you disagree with and contribute what you think makes your approach better. At the level of each professional in design or construction practice, I guarantee that the process of collecting, organizing and writing it all down will be satisfying for you as an experienced professional and for your successors who are desperate for the knowledge and experience that has allowed you to create the built environment.

I also challenge younger readers to contribute. There are no dumb questions in design and construction – asking is still how I learn.

Perhaps the best way for a younger reader to ask is (you guessed it) to share a story. A tale can just as well end with "What went wrong?" or "How can I do better next time?" All good.

If you are shy about sharing a story in social media, attach it to the relevant Procedure or i-Action in the sample QW project you have been given access to.

However you do it, please contribute your own tales. I have greatly enjoyed writing down tales that share and teach a bit. You will, too.

- Brian Palmquist, March 2015

Sharing Tales & Collaborating

Throughout this e-book I have pointed the reader to a handful of external websites and a number of detailed procedures, forms and templates that reside in my Quality-Works. net (QW) universe. As mentioned previously, any reader is invited to partake in QW's knowledge and experience through the venue of a sample project to which readers are given access. I expect most readers will find some new knowledge or experience that assists their practice of design and construction. But I can also guarantee that "your model may differ."

To gain access to the QW sample project, where examples are housed, you will first need to go to http://quality-works.net/signup.php

Read the QW subscription agreement, which describes the terms and conditions for your free access to the one or more sample projects connected to this e-book (these terms are typical for access to a sample project in any web-based application). If you agree, tick the "I have read and agree..." box, then click Next.

Sign up as a Student Free Subscriber. This will give you free access to the shared sample architectural project.

Once your subscription has been accepted (may take a day or two as I review each personally), you will receive a username and password, which you can use to access the Quality-Works.net (QW) application at

https://Home.quality-works.net

When you have signed in you will see one or more full sample projects. Each tale in this e-book may also have one or more hyperlinks that will take you to a specific location in the sample projects. If you are not logged into QW, you will need to do so before you can go to linked locations in QW projects.

As much as I have enjoyed writing AAGC, I would be that much happier if it was to become a resource for the discovery, debate and refinement of design and construction best practices. I have structured the "AAGC Sample architectural project" to support that goal.

When you login to the sample project as described in the Introduction, you will see a home page. Click on "Work Plan" to the right and you will arrive at the AAGC Project Work Plan (PWP) for the sample project:

Figure 193 - QW Login Screen Top

To see the array of comments, forms and templates opened to date, click on "Procedure Detail" center left (above).

Figure 194 - Top Portion of Sample PWP

To open a specific Procedure in the PWP, click on its name, in this case, "Assemble Project Staff":

Figure 195 - PWP Details

semble Project Team				
	Assemble Project staff			
	Procedure Notes			
	2014/03/24 - Read Write Reader can enter diary comments			
	Actions added for this procedure			
	Create New Action			

Number	Item Name	Due Date	Created	Closed Floor Status	Priority	Type - Subtype	Assigned To	
A00061 R0	Read Write Contact functionality		2014/03/24	Open	11 Reference	Information	Brian Palmquist	Edit Copy Upload

Figure 196 - Click on Assemble Project Staff

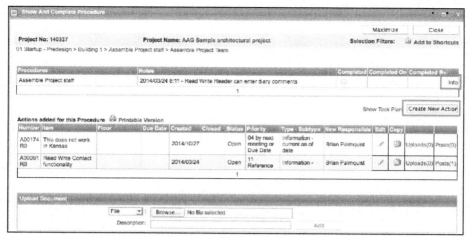

Figure 197 - Procedure Dialog Box

Your login rights allow you to add i-Actions (a.k.a. "Create new Action"), also open and edit or add Posts to i-A's.

If you wish to see the instructions for a particular Procedure, click on "Info" to the right of the Show and Complete Procedure box and printable work instructions applicable to the Procedure will appear:

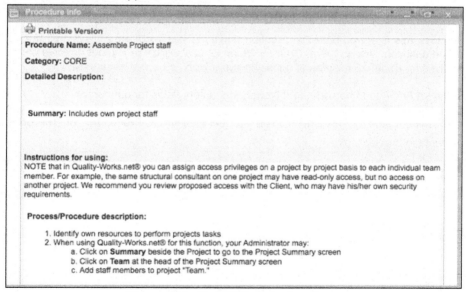

Figure 198 - "Info" Work Instructions

Back to the Show and Complete Procedure box. To create a new i-A about the Procedure, such as your own experience or a suggested modification to the Procedure or the AAGC book, click "Create New Action", which will open the "Add Action" i-A creation box:

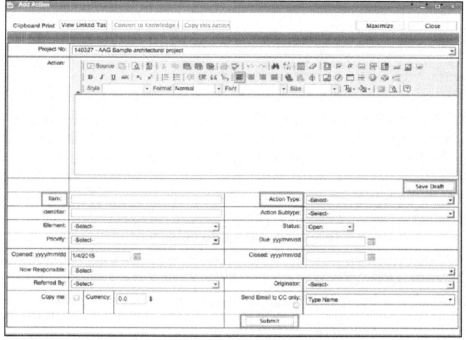

Figure 199 - Create i-A

Because this is a web-based application, anything you add from here on will not be saved until you click on "Save Draft" or "Submit." QW will close you out (to save bandwidth) after about 15 minutes of not saving your work, so do so frequently.

In the i-A Add Action box are the following mandatory fields for data entry:

Item is your name for the item, could be "A better idea for this Procedure", or "This does not work in Kansas", etc.

Action Type is just that, selected from a drop down list of Options:

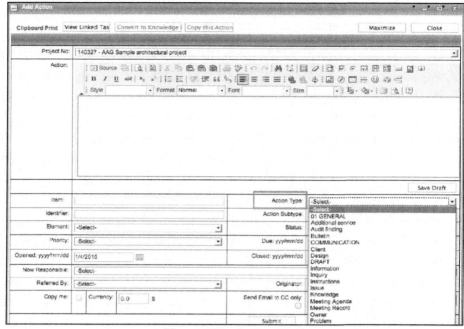

Figure 200 - Action Type

After you select the Action Type, there may be a short pause as QW loads the Action Subtypes associated with the selected Action type. Note that every drop down or pick list in QW is subscriber editable. In the case of the sample project, you are stuck with my lists.

Action Subtype is a subheading of the type of Action:

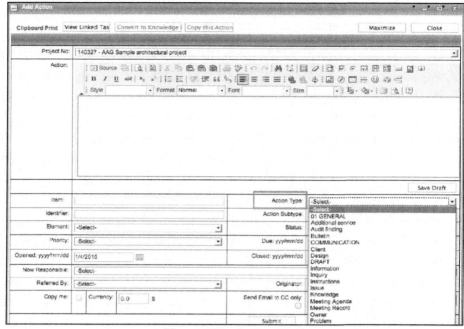

Figure 201 - i-A Subtype

Priority is a relative scale from "Urgent" to "Reference" that helps prioritize. When in doubt, select "by next meeting or Due Date" so that you can assign a due date:

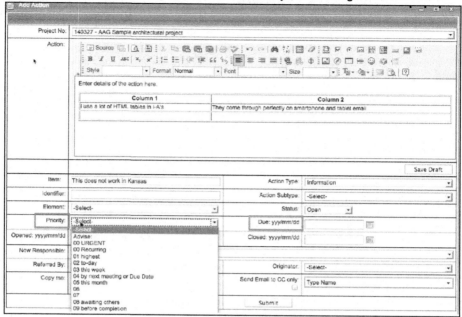

Figure 202 - i-A Priority & Due Date

You can select a Due Date, but this is not mandatory.

Now Responsible is the person you are assigning the i-A to. If you are commenting about the e-book and want my response, make me "Now Responsible" so I will get your message.

Figure 203 - i-A Now Responsible

Notice that the names of all of the persons with access to this project are in this list, which may get quite long over time (remember the 800 on my biggest project) – thanks for "auto fill", which allows you to start typing the assignee's first name, to drop to that name on the list. Also note that in contractor mode, names are organized alphabetically by first name, which I love as it moves "Brian" way above where I would be with "Palmquist."

The names of all readers of AAGC who request access to the sample project will be visible in this list, also in the next list.

Referred By is you, the creator and sender of the inquiry or suggestion.

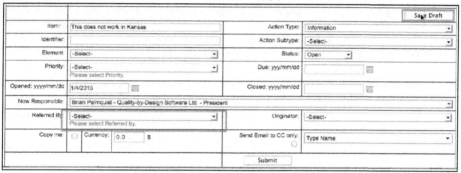

Figure 204 - i-A Referred by

When you have entered this data, you can select "Save Draft" and QW will upload info to date to the server. If you have missed something you will be advised.

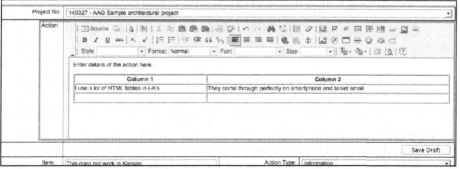

Figure 205 - i-A Missed "Referred By" and "Priority"

Before seeing what happens when/after you upload information, let's look at the optional fields that may be filled in.

Action is where we describe the details of the i-A. If we have opened a template, that will pre-populate the dialog box. If not we can simply enter our own data:

Figure 206 - Action details in i-A

There are a large number of editing tools in the beige toolbar that can enrich the "Action" data field with tables, different fonts and colors, etc.

Sometimes (a sprinkler system is a good example) there are "Elements" being consid ered that are systems spread out in a building. The Elements list allows such systems to be identified in as little or as much detail as has been set in the Elements list:

Item:	This does not work in Kansas	Action Type:	Information
Identifier:		Action Subtype:	-Select-
Element:	-Select-	Status:	Open
	03 Rebar		
Priority:	04 Masonry	Due: yyy/mm/dd	
	05 Metals		
Opened: yyy/mm/dd	05 Structural steel	Closed: yyy/mm/dd	
	05 Steel deck		

Figure 207 - i-A Elements

If you wish to reference something like a Consultant report number, a meeting minutes number, etc., the free form "Identifier" data field serves that purpose:

Item:	This does not work in Kansas	Action Type:	Information
Identifier:	THG #45-6	Action Subtype:	-Select-
Element:	05 Structural steel	Status:	Open
Priority:	11 Reference	Due: yyy/mm/dd	

Figure 208 - i-A Identifier

You can assign a Due Date – generally, a single reminder is sent to the "Now Responsible" and "Referred By" parties a day or so before the Due Date – just one reminder, in the spirit of not hectoring. And when the i-A is closed you can record that date as well, and decide whom you wish to notify that you are "done".

To upload the i-A to the i-WP, select "Save Draft". After a few seconds, the data is uploaded and a unique Tracking Number is assigned. The "R00" means "Revision 0":

	+		Save Draft
Item: A00186 R00	This does not work in Kansas	Action Type:	Information
Identifier:	THG #45-6	Action Subtype:	-Select-
Element:	05 Structural steel	Status:	Open

Figure 209 - The i-A has been Assigned #A00186

Now that the item is "in" QW, you can continue to work on it. When you are ready to communicate it, select "Submit" at the lower right corner. You will then be asked if you wish to email the item to the "Now Responsible" person:

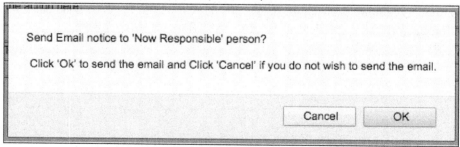

Figure 210 - Email i-A?

Sometimes you are simply creating a record, not sending it to anyone, in which case you will select "Cancel". Otherwise, select "OK" and the assignment is on its way. If you have made me "Now Responsible" and want me to receive your comments, you will need to click "OK."

Before the i-A email has arrived, it will be lodged with the Procedure where you started:

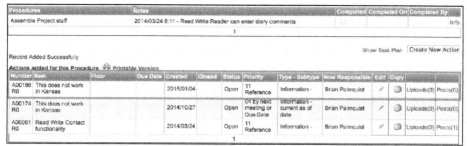

Figure 211 - i-A as Record

The balance of QW i-A functions is fairly typical of web-based applications:

If you wish to attach an image to the i-A, look for the "Uploads" selection either during draft stage or after you have sent the item. A sent item can be reopened and items attached, but it cannot generally be edited as to content, to preserve the efficacy of the project's records. Administrators can reopen i-A's, which causes the revision number to advance, but you will not be able to do that in the sample project.

The i-A email you receive will look something like this (this one is from an iPad):

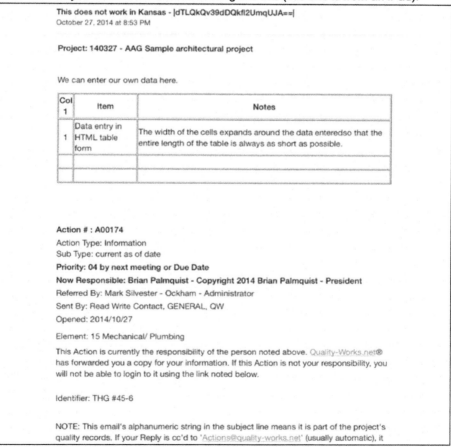

Figure 212 - i-A as Email on iPad

There are a few oddities about this email. The odd alphanumeric string by the subject line helps QW find the originating Action from among the thousands in QW. This ensures that in addition to being treated like standard email (i.e., shows up on your computer, tablet or smartphone), any responses to the i-A, including attachments, third party comments, etc., are all captured to the originating i-A. Yes, records are captured automatically! Even better, the captured record is limited to new information, rather than strings of strings of strings of preceding email.

If I decide to "reply all" to this i-A from my iPad, the reply has automatically included the alpha string plus the email address actions@quality-works.net. So long as this email address and the correct alpha string are on 2nd, 3rd, 4th generation emails, they and their attachments will all be captured to QW.

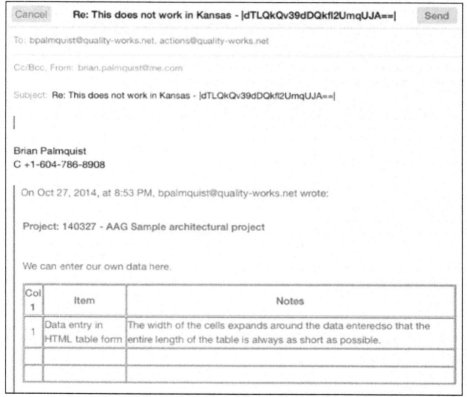

Figure 213 - Reply to i-A from iPad

Reviewing Construction the Simplest Way

I like to think of myself as somewhat "tech-savvy", but every once in awhile I get blown away when I discover something that moves the goal posts. Such was my recent discovery about checklists, in fact any HTML table document included in QW or other web-based applications.

This tech "tale" starts with my frustration that my own QW's HTML editor was unable to deal with tick boxes – you know, the little square boxes that one would like to tick off indicating successful completion of something. I was wrapped up in Atul Gawande's approach to checklists (more about that in an earlier tale) and the graphic simplicity of a simple list of items with tick boxes. But QW was unable to handle them in the way I wished. Try as I might, ticking off a box, then emailing it, resulted in the ticks being undone somewhere in the ether. Nuts.

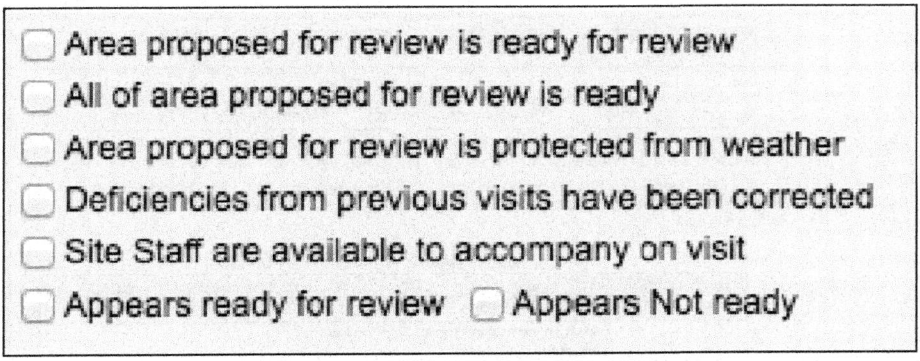

☐ Area proposed for review is ready for review

☐ All of area proposed for review is ready

☐ Area proposed for review is protected from weather

☐ Deficiencies from previous visits have been corrected

☐ Site Staff are available to accompany on visit

☐ Appears ready for review ☐ Appears Not ready

Figure 214 - Sample Checklist Tick Boxes

I am sure there is a tech answer to why this does not work in an HTML/email/web environment, but it's beyond me. So I returned reluctantly to the table type documents I have been using to set up checklists and a wide variety of other forms in QW.

Another problem: those forms do not transfer to the offline version of QW developed for my iPad. My software guru says this is possible to program for a 5-figure dollar investment that I don't have.

So I returned to the table format I had previously been using with success:

An Architect's Guide

Issuing company :

Posts for Action - 08 Contractor REVIEW Aluminum Entry Doors

Action		Posted: 2014/09/16 06:16:32 PM
Now Responsible:	Read Write Reader, GENERAL,	Referred By: Read Write Reader, GENERAL,

Project No:	131218 Comprehensive project	Number:	A00098 R0	Item:	08 Contractor REVIEW Aluminum Entry Doors
Priority:	11 Reference	Status:	Open	Type:	Sample
Due:	12:00:00 AM	Opened:	2014/09/16	Closed:	12:00:00 AM
				Element:	
				Subtype:	

Description: To reply to this email, select "Reply All", fill in the information indicated and select "Send".

Weather: ☐ Sunny ☐ Cloudy ☐ Mixed ☐ Rain showers ☐ Rain ☐ Snow ☐
Other/details:

Ref. Dwgs: | Grid Refs:

Report date yymmdd: | Review date yymmdd: | Associated Quality Report #:

Below is a summary of work reviewed. In case of deficiency or non-conformance, fill in details below in the "Comments/ Sign Offs" column LEGEND: ☑ or "Y" = item reviewed and accepted; X = item reviewed and unacceptable; items not applicable at this time are so marked; if you are using an iPhone or iPad you can select a "COMMENT/ SIGN OFFS" text box and dictate comments.

Contractor Review of Aluminum Entry Doors

Done	N/A	DESCRIPTION	COMMENTS / SIGN OFFS:
		Before construction of this trade/area starts:	
		A SQP – Subcontractor Quality Plan(s) submitted & accepted?	
		Mockup requirements identified & scheduled? ? "First of" installations reviewed by Consultant	Includes construction mockups that may not have been specified, based on designer or builder experience.
		Submittals: ? Scheduled? ? Submitted? ?	

Figure 215 - Checklist in HTML Table Format

I already knew that HTML tables translate well to smartphones and tablet computers. How best to incorporate them in a project so that a consultant or project manager in the field or a superintendent can make effective use of them?

Ideally, a person in the field should be able to use the standard email reader/writer of a smartphone or tablet to access a form, template or checklist for a project, open a copy, fill it in and send it to whomever – consultant, owner, contractor or subcontractor. And here's the trick – even though the form is opened from, say, a smartphone, when it is executed and issued your i-WP should grab the document and automatically append it to the issuing project, with full details of what it said and who it was sent to. And the i-WP should capture all the follow through until items raised in the form are closed.

Through some trial and error, I actually got this to work with the following sequence:

When you set up a project i-WP in QW it includes a wide variety of forms, templates and checklists that may be needed for the project. Let's say you are the superintendent, so you are focused on checklists and non-conformance/deficiency forms. Simply open a sample of these within the project i-WP, assign it to yourself (the superintendent) and issue/submit it to yourself blank. Or you can get a project manager or coordinator to do it for you, so long as it ends up on your smartphone or tablet.

The superintendent will receive the forms and checklists on his/her email reader, i.e., smartphone or tablet or laptop – in fact, all of the above. The recipient sets up an Outlook (or similar application) email folder for the project and moves the received emails into that folder. No rocket science here.

▼ 00 MOCKUP REQUIREMENT IDENTIFICATION – |RD2KKR91PBSQ86PY06N/TG==|	
Brian Palmquist	14-09-18
00 Mockup Requirement Identification – |rd2Kkr91PbSQ86py06N/tg==|	
▼ 00 QUALITY CHECKLIST TEMPLATE FILLED	
Brian Palmquist	14-03-03
00 QUALITY CHECKLIST TEMPLATE FILLED	
▼ 00 SUBMITTAL REVIEW – |KDU2JXB3Z00JUR/EBRBP4W==|	
Brian Palmquist	14-09-18
Re: 00 Submittal Review – |kdU2jXB3Z00juR/EbRBP4w==|	
▼ 00 SUBMITTAL REVIEW – ALTERNATIVE OR SUBSTITUTION – |/MA990QMJMIL6VJL6S4/FG=	
Brian Palmquist	14-09-18
00 Submittal Review – Alternative or Substitution – |/mA990QmjmIL6VJl6s4/fg==|	

Figure 216 - Checklists in Outlook Email Folder

Then, next time the superintendent needs, say, an aluminum entry door checklist, she/he uses the smartphone, tablet or laptop to navigate to the folder, selects the form and then selects "Reply All".

●○○○○ TELUS 🛜   6:55 PM   36% 🔋

❮ All Inboxes (10)   ⌃ ⌄

Contractor Review of Aluminum Entry Doors

Done	N/A	DESCRIPTION	COMMENTS / SIGN OFFS:
		Before construction of this trade/area starts:	
X		A SQP – Subcontractor Quality Plan(s) submitted & accepted?	
X		Mockup requirements identified & scheduled? ? "First of" installations reviewed by	Includes construction mockups that may not have been specified, based on designer or

Figure 217 - "Reply All" is at the Bottom of the iPhone Screen

The user now has a form that she/he can complete in the field and email, selecting appropriate recipients. Because the copy of the form on the hardware device originated from the QW project, all copies issued from it are automatically captured by QW to the originating location, as well as responses to the issued email.

A complete record of who said what to whom, including any photos or documents attached, is automatically aggregated to the issuing document.

Where a review is interrupted or delayed, standard smartphone functionality generally offers the choice of saving a draft for however long is necessary:

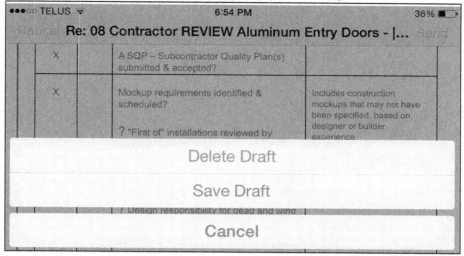

Figure 218 - Save Draft Preserves an Incomplete Document on an iPhone

This means, for example, that where a cladding review is interrupted or takes place over an extended time period, the "draft" is preserved on the smartphone/tablet until ready for completion and issue. It also means that if a checklist is used ten times on a project, all ten of those issuances are individually and automatically captured in the i-WP with the originating checklist, as well as every response, additional document, etc., all in chronological order. So all of the cladding reviews, etc. are captured to one place in the project with no additional effort by any of the project team members.

One cautionary note: until issued from the originating smartphone or tablet, a draft form has not yet been saved. For this and other reasons I recommend frequent backups of smartphones and tablets.

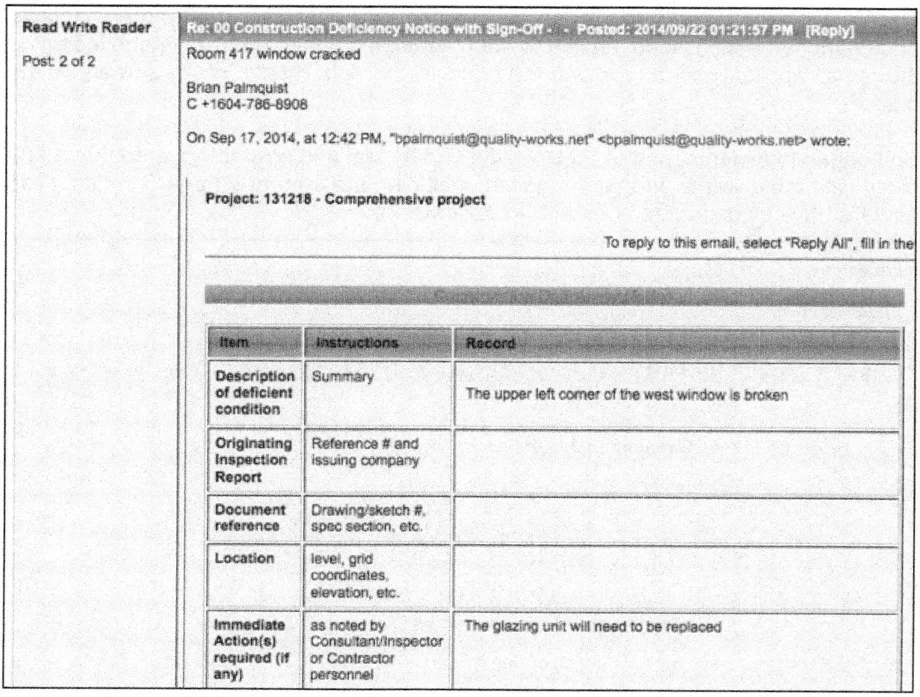

Re: 00 Construction Deficiency Notice with Sign-Off - - Posted: 2014/09/22 01:21:57 PM [Reply]

Room 417 window cracked

Brian Palmquist
C +1604-786-8908

On Sep 17, 2014, at 12:42 PM, "bpalmquist@quality-works.net" <bpalmquist@quality-works.net> wrote:

Project: 131218 - Comprehensive project

To reply to this email, select "Reply All", fill in the

Item	Instructions	Record
Description of deficient condition	Summary	The upper left corner of the west window is broken
Originating Inspection Report	Reference # and issuing company	
Document reference	Drawing/sketch #, spec section, etc.	
Location	level, grid coordinates, elevation, etc.	
Immediate Action(s) required (if any)	as noted by Consultant/Inspector or Contractor personnel	The glazing unit will need to be replaced

Figure 219 - Checklist Automatically Captured in Post to Original

But, I thought, was the result actually usable? I selected the smallest likely recording device, an iPhone, and proceeded. To my delight everything worked as discussed above, with the added benefit that the table cells gave me a large target to tap my finger in before making a comment.

I was even more delighted to discover that the voice recognition software in the iPhone and iPad (and I suspect other hardware) allowed me to speak my comments, etc., and have them recorded with minimal extra typing or translating.

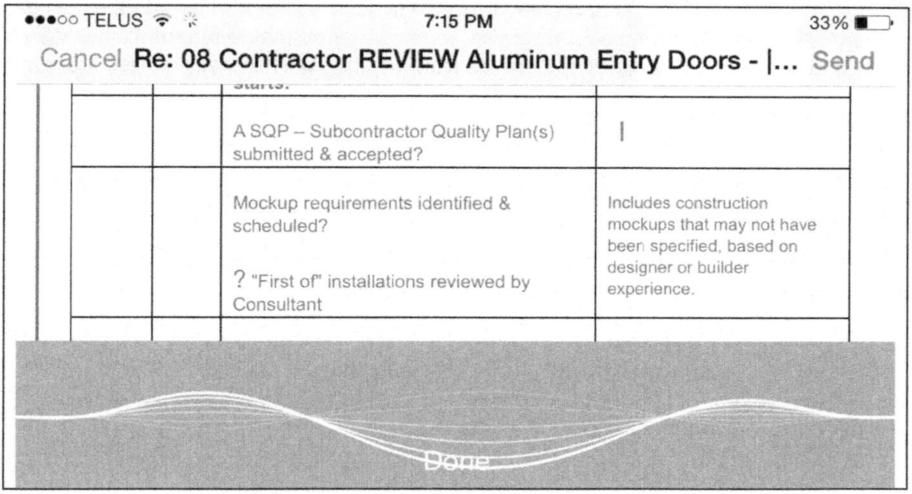

●●●○○ TELUS 📶 ☀   7:15 PM   33% 🔋

Cancel **Re: 08 Contractor REVIEW Aluminum Entry Doors - |...** Send

		starts:	
		A SQP – Subcontractor Quality Plan(s) submitted & accepted?	I
		Mockup requirements identified & scheduled?	Includes construction mockups that may not have been specified, based on designer or builder experience.
		? "First of" installations reviewed by Consultant	

Done

Figure 220 - Sound Dictaion into Checklist

Once I had determined that this functionality works, I noticed that after a HTML table document had been issued, replied to, then again and again, pretty soon the indenting that email applications do bumped the table off the right margin of the screen, or the printer.

So I began to experiment with table widths and by trial and error discovered that a 500 pixel wide table works on every medium, including the current iPhone 5 series. (The new 6's have bigger screens so will naturally work).

From iPhone

Brian Palmquist
C +1604-786-8908

On Sep 16, 2014, at 11:51 AM, "bpalmquist@quality-works.net" <bpalmquist@quality-works.net> wrote:

Project: 131218 - Comprehensive project

400 pixels			
450 pixels			
500 pixels	this is max showing up first time on iPhone		
550 pixels			
	third time from iPad		

Figure 221 - Experimenting with Table Width

So at the end of the day we have the following: a standard width of 500 pixels for any HTML form that we want to manage from an iPhone or similar smartphone; a list of project-specific forms, templates and checklists that lives in a standard email folder on the hardware device; and an emailed report of each issuance that is automatically stored where it should be, including all responses from wherever. The user can use standard email procedures, just "Reply All" for each response. And any table cell can be filled in by speaking to the hardware device. To recap:

1. Set up an i-WorkPlan (i-WP) for your project
2. Open a blank copy of forms, templates and checklists as/when needed
3. Email the blanks to the person(s) who will use them
4. Place them in a folder on the person's smartphone/tablet with the job name – it will appear the same on all interconnected hardware devices.
5. Each time you wish to use a copy of the form, open it from the email folder and "Reply All", adding new recipients as required.
6. Fill in, using voice recognition if desired and available.
7. Send when complete. When responses come, manage them as ordinary email except always "Reply All" – QW does the rest.

Is that cool or what?

In "Down Detour Road", author Eric Cesal does a much better job than I in expressing his annoyance about the hijacking of the word "architect", largely by the software industry:

"It occurred to me that over the last generation, while a bunch of smart people anguished over the distinction between "architect" and "designer" and between "intern architect" and "interior architect", someone stole our damn name." [1]

In the interests of clarity I have used the following terms in this book (items in "quotes" are from wikipedia.org):

AAGC is short for "An Architect's Guide to Construction".

Addendum is a document issued during the course of bidding, for the purpose of clarifying the intent of the bid documents, or identifying specific products or systems that may be approved as acceptable alternatives.

Additional service should be one of your favorite phrases and refers to something in addition to your basic professional design or construction services, which you should be paid extra for.

Architect refers to those involved in building architecture, who have become registered or chartered in one or more jurisdictions, typically countries, provinces or states. Also called "registered architect." It excludes "software architect", "solutions architect", etc.

BIM – Building Information Modeling "is a process involving the generation and management of digital representations of physical and functional characteristics of places."

Builders refers to the general class of businesses engaged in the construction of buildings; includes general contractors, construction managers, subcontractors, suppliers and installers.

Client refers to the person or company who has engaged Designers and Builders to design and construct a building.

Consultant is an individual or company applying specific expertise to the design and construction of a building, which is what many of you are, together with the other folks also called "Designers."

Contractor refers to the general class of businesses engaged in managing the construction of buildings; includes general contractors, construction managers, etc.

Construction Manager refers to a business engaged in the construction (and sometimes design) of buildings by developing and managing project teams and budgets, schedules and work scopes on behalf of clients.

1   Cesal, Eric J., Down Detour Road: An Architect in Search of Practice, MIT Press, Cambridge, MA, USA, 2010, p. 17

Designers refers to the group of professions involved in the design of buildings, i.e., architects, landscape architects, structural, mechanical and electrical engineers, etc.

HTML stands for Hyper Text Markup Language, the standard language used to create web pages. A web browser, including a smartphone or tablet computer, can read and mark up HTML documents or forms. HTML is used to create Quality-Works' structured documents, including i-Actions and i-WorkPlans.

IFC – Issued for Construction documents are those that have been formally issued to the builder for the construction of the project, different from design documents, bid/tender documents, permit documents, etc.

Intern architect refers to an architectural education program graduate or foreign trained architect involved in a formal pre-registration program leading to designation as an Architect in a particular jurisdiction.

Mockup is a sample of a proposed piece of the building design, executed in advance of the commencement of the balance of that work so that the builders can understand what has been designed and the designers and clients can confirm the sample of work meets the intent of the design documents.

P3 – Public Private Partnership is a form of project development wherein a consortium of designers, builders, financiers and operators designs, builds, finances and operates a facility for a specified time period and for a specified fee.

PDF or Portable Document Format "is a file format used to present documents in a manner independent of application software, hardware, and operating systems"

Post Tender Addendum is a clarification to the bid documents issued after tenders have been received. This most often occurs when the bidder nominated to receive the construction contract concludes negotiations with the client and consultants, and it is desirable to include those results in the IFC documents.

RFI – Request for Information occurs when a contractor asks a designer for information about an aspect of the construction that appears unclear to the contractor. RFI's may include information not referenced, information thought to be missing, information thought to be wrong, etc. RFI's require a formal response from the addressee.

Student Architect refers to a student in an architectural program.

Submittal is anything a Contractor provides to a Consultant to confirm that what will actually be built is what was designed. Submittals may include samples, shop drawings, manufacturer's literature, warranties, etc. Some Contractors consider mockups to be a form of submittal.

Other terms are defined in the body of the book.

CPSIA information can be obtained
at www.ICGtesting.com
Printed in the USA
LVOW13s0928040518
575929LV00004B/18/P